The Lived Experience of Climate Change

Dina Abbott · Gordon Wilson

The Lived Experience of Climate Change

Knowledge, Science and Public Action

Springer

Dina Abbott
Development Geography
University of Derby
Derby
UK

Gordon Wilson
Environment and Development
Open University
Milton Keynes
UK

ISBN 978-3-319-17944-5 ISBN 978-3-319-17945-2 (eBook)
DOI 10.1007/978-3-319-17945-2

Library of Congress Control Number: 2015937014

Springer Cham Heidelberg New York Dordrecht London

Printed on acid-free paper

Springer International Publishing AG Switzerland is part of Springer Science+Business Media (www.springer.com)

Preface

On a grey December morning in 2007, one of the authors of this book—Gordon Wilson—parked his folding bicycle at the Catholic University of Leuven in Belgium to join a 'Task Force on Sustainable Development' that was supported by the European Association of Distance Teaching Universities (EADTU). The meeting comprised university lecturers from five European universities. It discussed possibilities for partnership on a climate change curriculum.

Through several months of deliberation, none of the Task Force members saw any point in replicating university curriculum on climate change that already existed across Europe at that time. While not extensive in its reach, this curriculum tended to be led by the natural science, and did so perfectly adequately. Thus, through a joint desire to do something new, the idea of the lived experience of climate change—the everyday experience of individuals and groups of climate-related events such as drought, floods and sea level rise—was born. The aim was not to deny or attack the science but to add a very different, complementary knowledge in order to develop a more holistic approach to understanding the climate challenge.

Gordon Wilson suggested the concept of lived experience to the Task Force because of its use in research on the underlying dimensions of poverty in developing countries, which was his subject area. His 'reward' for doing so was to be charged with producing a concept note which was presented to the EADTU annual conference in September 2008. It was well received and with further EADTU support a successful bid was made to the EU Erasmus programme to produce Masters curriculum as open educational resources on the lived experience of climate change. The project also committed to piloting the curriculum to students across our universities in an exercise of virtual mobility (that is, using social media for students and lecturers to transcend individual, institutional boundaries). The project involved eight universities (five distance teaching and two conventional face-to-face) across six EU countries plus the EADTU. It started in October 2009. It is described in this book in Chap. 6 Box 6.2.

The project, as judged by external assessors and 26 students who participated in the pilot delivery, was a great success. Meanwhile, in producing the teaching resources, we had been forced to think much more deeply about the fledgling core concept of lived experience. That was only the start, however, because then the students challenged us via social media tutorials, structured group work and their essays to go further still.

By the time the project ended in April 2012, everyone who had been associated with it felt that, in the concept of the lived experience of climate change, there was something to tell the world. We—Gordon Wilson and Dina Abbott who between us had produced most of the written teaching texts and workbooks—took up the challenge, one result of which is this book.

Of course, as we have engaged ever deeper with the book's conception and the writing of it, we have learned much more about the lived experience of climate change, its nuances, its explanatory powers, its contradictions and its possible use to inform policy. We are continually reminded, however, that we owe an enormous debt to our colleagues on the original project and the students who participated so enthusiastically in the pilot. Literally, this book could not have happened without you.

Contents

1 **Introduction: A Wealth of Lived Experience** 1
- 1.1 Introduction 1
- 1.2 Lived Experiences of Weather, Floods and Climate 2
- 1.3 The Rich Diversity of Lived Experience 5
  - 1.3.1 Contrasting Lived Experiences of Rich and Poor 6
  - 1.3.2 The Lived Experience of Public Attempts at Climate Change Mitigation and Adaptation in an Affluent Country 7
  - 1.3.3 Connecting Climate Change and Climate Change Mitigation to Lived Experience of Vulnerability of Tropical Forest Communities in East Africa 9
  - 1.3.4 The Lived Experience of Environmental Activists 15
- 1.4 From Flood Victims in the United Kingdom to Small Farmers and Drought-Resistant Food Grains in Zimbabwe: Making Sense of the Lived Experience of Climate Change 20
- 1.5 The Building Blocks of This Book 24
- References 24

2 **Exploring the Lived Experience of Climate Change** 27
- 2.1 Introduction: Lived Experiences as Rich, Complex Narratives 27
- 2.2 Lived Experience as a Social Process of Making 30
- 2.3 The Complex Interaction of Structure (Power Relations) and Agency that Makes Lived Experience 33
- 2.4 Lived Experience in Relation to Interacting Phenomena 38
- 2.5 Lived Experience in Relation to Culture and Value 41
- 2.6 The Lived Experience of Climate Change: Personal and Collective 43
- 2.7 Lived Experience and Scale 45
- References 48

3 Lived Experience and Scientific Knowledge of Climate Change 51
3.1 Introduction 51
3.2 Science and Lived Experience: Whose Reality? 55
3.3 Diverse Scientific Perspectives and the Political Exploitation of Diversity 67
3.4 Working with Scientists 70
3.4.1 Working with Physical Scientists: A Social Science Story 70
References 74

4 Representing Climate Change: Science, Social Science, Interdisciplinary Approaches and Lived Experience 77
4.1 Introduction 77
4.2 Climate Change as a 'Wicked Problem' 79
4.3 Fundamentals of Natural Science Inquiry and the 'Scientific Method' 81
4.4 Underlying Assumptions of the 'Scientific Method' and Positivist Objectivity 84
4.5 Fundamentals of Social Science Inquiry and Power Relations 88
4.5.1 A Shift from Numbers to Qualitative Data 90
4.5.2 Underlying Ontological Assumptions and Celebration of Subjectivity 91
4.6 Interdisciplinarity, Representation and Climate Change 95
4.6.1 The Politics of Representation 98
References 102

5 Lived Experience and the Advocates of Local Knowledge 105
5.1 Introduction 105
5.2 Local Knowledge and What Works 110
5.2.1 Sharing or Extracting Knowledge Through Participatory Processes: The Contribution of Development Studies 113
5.2.2 Local Knowledge, Power and Participatory Processes in Environmental Studies 115
5.2.3 Local Knowledge and the Climate Change Literature 117
5.3 Local/Indigenous Knowledge and Identity 119
5.4 Local Knowledge, Indigenous Knowledge and Lived Experience 121
References 124

6 Lived Experience and Discourses of Mitigation, Adaptation 127
6.1 Introduction 127
6.2 Mitigation and Adaptation and the IPCC 128

6.3 Adaptation and Mitigation in the 4th Assessment Report (2007) . . . 130
6.3.1 Adaptation as Covered by the 2007 Report of WGII . . . 130
6.3.2 Mitigation as Covered by the 2007 Report of WGIII . . . 132
6.4 Adaptation and Mitigation in the 5th Assessment Report (2014) . . . 133
6.4.1 The 2014 WGII Report 'Impacts, Adaptation and Vulnerability' . . . 133
6.4.2 The 2014 WGIII Report 'Mitigation of Climate Change' . . . 135
6.5 The Lived Experience of Mitigation and Adaptation . . . 140
References . . . 153

7 Lived Experience and Engagement on Climate Change . . . 155
7.1 Introduction . . . 155
7.2 The Public Will to Know and Act as a Problem of Science Communication . . . 157
7.3 The Public Will to Know and Act, and Disagreement Over Climate Change . . . 160
7.4 The Public Will to Know and Act, Socially Organised Denial and Lived Experience . . . 168
References . . . 172

8 Lived Experience, Science and a Social Imagination . . . 175
8.1 Introduction . . . 175
8.2 To What Extent Is It Desirable to Expand the Boundaries of Our Lived Experiences Beyond Prevailing Social Norms of Attention? . . . 178
8.2.1 Communicative Action and Hermeneutic and Emancipatory Learning . . . 181
8.2.2 The Social Imagination . . . 184
8.3 To What Extent Is It Desirable to Expand Scientific Knowledges About Climate Change Beyond Their Boundaries? . . . 186
8.4 How Might Expansion of Climate Change Knowledge Be Enabled and Who Might Enable It? . . . 190
References . . . 197

9 Rationalist and Public Action Theories of Knowledge in Climate Change Debates . . . 201
9.1 Introduction . . . 201
9.2 Towards a Targeted Climate Change Policy . . . 203
9.3 Positive and Normative Assertions of Policy Making . . . 205
9.4 The Role of the Media . . . 207
9.5 The Rationalist Model of Policy Making . . . 209

9.6 The Social Process of a Non-linear Public Action Approach and Lived Experience. . . . 211
9.7 Power Politics and Contested Knowledge 215
9.8 Lived Experience and Public Action Through Demonstration. . . 217
References 219

10 Institutionalising Lived Experience in the Public Policy Process 223
10.1 Introduction 224
10.2 Why Should We Wish to Institutionalise Lived Experience in Public Policy Debates About Climate Change? 225
10.2.1 Government and Its Local Institutions 225
10.2.2 The Private Sector 228
10.2.3 Scientists Across the Social and Natural Divides. 229
10.2.4 Engineers 230
10.2.5 Civil Society 231
10.2.6 Us, the Authors of This Book. 232
10.3 The Issues (and Problems) Associated with Institutionalising Lived Experience in Public Policy Debates About Climate Change 236
10.3.1 'Getting Real' 236
10.3.2 Creating the Conditions for Productive Engagement with Lived Experiences, and Between Lived Experiences and Other Forms of Knowledge. 237
10.3.3 The Issue of Power 239
10.4 Ways of Institutionalising. 242
10.4.1 Institutionalisation Through Monitoring Independent Social Media 243
10.4.2 Direct Governmental Institutionalisation Through Focus Groups 245
10.4.3 Independent Institutionalisation Through Environmental Action Groups 246
10.4.4 Bringing It Together: Institutionalising as Space Making. 250
References 251

11 A Public Action Approach to Knowledge and Intervention to Meet the Climate Challenge 255
11.1 Introduction 255
11.2 A Public Action Approach to Knowledge 257
11.3 Public Action, Knowledge and Engagement on Climate Change: The Broader Picture. 265
References 268

Chapter 1
Introduction: A Wealth of Lived Experience

Abstract We need to understand better the societal responses to the idea and reality of climate change. Analysis of lived experiences enables us to embark on that task. Lived experiences, however, even when we try to make sense of them as collective experiences, are extremely diverse. This chapter describes and analyses the experiences of flood victims in affluent countries, vulnerable forest communities in two poor countries, opposition to mitigation attempts, and two environmental activists. From these it creates a conceptualisation that is based on the interplay of broad contextual influences, proximate influences resulting from climate-related events, and the human capacity to reflect and learn from action and engagement with one another. It ends by providing three building blocks for the book as a whole: the conceptualisation that it has generated, the comparison with dominant scientific accounts of climate change, and the challenge of public engagement, action and policy for intervention.

1.1 Introduction

This book is driven by the need to rethink radically our understanding of societal responses to climate change, the knowledge that informs and influences those responses, and the consequent implications for policy makers and practitioners.

In meeting this need, the book places everyday lived experience as central to the climate change challenge. We, the authors, argue that understanding and sharing our diverse lived experiences as citizens across the world, sharing them with the sciences, and sharing the experience of intervention practice are essential for developing a more complete knowledge. Through this more complete knowledge, citizens have the potential to reclaim the power to influence public agendas and policy on climate change.

The book, therefore, represents a comprehensive and powerful lens on climate change where we and you, the readers, are about to embark on a journey which will sometimes take us into uncharted territory. Along the way we will explore and

D. Abbott and G. Wilson, *The Lived Experience of Climate Change*,
DOI 10.1007/978-3-319-17945-2_1

often challenge fundamental assumptions about science, about local knowledge, about citizen participation in decision-making, and about the relationship between knowing and policy on climate change.

Reviewers of books usually like to see neat definitions of the key concept up front. University lecturers especially want to see them. That is what we tell our students embarking on assignments—the first rule in communication is to define what you are communicating. It is, however, very difficult to start with an abstract definition of 'lived experience of climate change' without seeming tautologous. For example:

> The lived experience of climate change is the sense we make of the changing climate through living our lives.

Apart from the tautology in this example, the definition seems like common sense, that is, it is obvious. Interesting things happen, however, when we observe different lived experiences and ask how they are made and how they evolve. By the end of this chapter we will have offered you, if not a neat definition, at least a clear meaning and sense of what we mean by the idea. It will be open to any reader to reflect on that meaning.

Let us now begin.

1.2 Lived Experiences of Weather, Floods and Climate

It rained in the United Kingdom between late December 2013 and early February 2014. In itself, that was not unusual for a country that is not renowned for dry weather and sunshine. Nor was it unusual that it rained heavily, accompanied by strong winds, over a 6-week period. This was, after all, the depths of the British winter when storms sweep in from the Atlantic Ocean.

It did not, however, only rain heavily with an occasional gale. It rained excessively, almost every day, and the winds reached storm force. The rain in fact was unprecedented and southern England bore the brunt. Large swathes were affected by serious coastal flooding from high tides while inland the ground could absorb no more water and rivers burst their banks. Thousands of homes were affected and the news was awash with personal stories of rescue, damage and coping.

Several months later in August 2014, while much of Britain enjoyed a warm, sunny August the likes of which are not often seen, the 'The Guardian' newspaper published interviews with eight people who had been adversely affected by these floods (Barton 2014). The interviewees reflected on the physical and mental disruptions to their lives. Some had not yet been able to return to their homes because of the physical damage, others indicated that their lives could never be the same, with one describing how what was once her 'paradise' had left her numb emotionally. There was also a suggestion from some that their homes were allowed to flood by the authorities in order to save the local town of 38,000 inhabitants from going under water.

Yet the interviews also indicated a high degree of stoicism and re-building, emotionally and physically. There were even positives, such as a greater sense of community among those who suffered together. One interviewee, whose small business was turned upside down as flood waters entered his premises, and who kept paying his six workers during the period when they produced nothing, suggested that they are now much stronger as a team through their joint efforts to keep the enterprise afloat.

The interviewees reflected further on the causes of the floods. Two of the eight specifically mentioned climate change/global warming. Other explanations concerned lack of dredging in the rivers and failure to maintain drains that are now blocked.

The newspaper blog associated with these interviews attracted 122 comments over the 33-hour period that it was open. Many of the comments were sympathetic, both to the ordeal that the interviewees had faced and the ways in which they had coped and are continuing to cope. A significant number, however, were unsympathetic. They asked loaded questions, the following being a summary:

- What do these people (the interviewees and others who were flooded) expect if they choose to live in houses that are built on flood plains? They have only themselves to blame.
- Why are those who are now renovating their homes doing nothing apparently to flood-proof them?
- The interviewees all appear to be affluent people who have the money to cope with this kind of thing. Why don't they sort out their own mess—there are many others who are poor and in greater need of daily help from public resources, flood or no flood?

Some contributors to the blog also debated the possibility of learning from other countries such as the Netherlands and building houses in flood-prone areas on stilts. Finally, they discussed the impacts of dredging and maintaining drains, the gist being that if you save one area from flooding through such activities, the water and the associated problem will only emerge in another.

******

What is the relevance for this book of the above digest of a British newspaper feature article? Most obviously it concerns the experiences of eight people who lived through an unprecedented event in their lives, an event that was connected to climate change by the media, by politicians, and by many of those who were affected directly. The interviews themselves are mostly in the form of direct quotation, with only some scene-setting and context provided by the journalist. They appear, therefore, as direct *lived experiences* of the floods. They also suggest that the articulation of these experiences is a process of reflection and communication, and in this case, a reflection and communication that is some months after the trigger event.

The process itself goes beyond simply describing what happened personally. It invariably tells also of the longer-term impacts on lives, both bad and good. It indicates the communality of the experience. The newspaper reader-blog, meanwhile, goes beyond simple one-way communication that the interviews provide in that there is engagement between the contributors, albeit at times rather crude. The engagement touches on policy debates, such as those around flood defences, around planning policy with respect to house-building, appropriate insurance, and of stimulating innovation with respect to flood-proofing homes. Such reflection, communication and engagement become not just the articulation of the lived experiences but the lived experiences themselves. We are the product of the stories we tell about ourselves and how these may or may not be modified through communication and engagement with others.

The interviews themselves must, however, be qualified in relation to the fundamental content of this book, which concerns climate change. We are the first to agree that the eight experiences of 'The Guardian' feature were of weather rather than climate. Weather and climate are not the same although they are obviously related. We give and analyse the definition of climate that is provided by the United Nations Inter-governmental Panel on Climate Change (IPCC) later in Chap. 6. Generally speaking, for us, however, climate and climate change, and the associated concepts of greenhouse gases and global warming, may be defined as:

> Broad patterns of precipitation, temperature and wind, and their seasonal fluctuations. These patterns are relatively stable over significant periods of time. Thus one speaks of a Mediterranean climate with its hot dry summers and, its short rainy winter; a tropical climate with always high temperatures, but up to two rainy seasons per year; a temperate climate such as in Britain where it is never normally too hot or too cold and where it may rain – but not usually with the intensity of a tropical storm - at any time.
>
> *Climate change* as a physical phenomenon (we consider it as also a social phenomenon throughout this book) refers to disruption of these broad patterns through natural or human (anthropogenic) influences. The latter derive from economic and domestic activities that produce and increase the concentrations of *greenhouse gases* in the atmosphere. The term refers to gases, primarily carbon dioxide, but there are others including water vapour, methane, nitrous oxide and ozone, that trap part of the energy that arrives from the sun through their particular energy absorption properties. This in turn results in increasing temperature of the Earth, known as *global warming*, which causes sea levels to rise due to water expansion and clouds to hold more water. Global warming, however, represents an average temperature increase for the world but it is unevenly distributed across land and sea. In other words, some areas and seas will warm faster and others more slowly, which effect changes in air and ocean currents that are direct drivers of the diversity of weather patterns across the globe.

The final sentence above brings us to the subject of weather which concerns precipitation, temperature and wind on a daily basis. One crude distinction between it and climate is in the expression, 'Climate is what you expect, weather is what you get'. Another is that climate is 'average weather', although a straight average will always be unsatisfactory because it hides the fluctuations, such as the extremes of temperature and precipitation that occur.

Was the extreme weather in the British winter of 2013 and 2014 connected to climate change? In the experience of many it was. Moreover, after a benign

Fig. 1.1 The flood plain of the River Severn on the Wales-England border does its job after heavy rain in December 2014 (G. Wilson, 2014)

summer and autumn, flooding from excessive rainfall returned a year later (Fig. 1.1). Experience, however, is not science which had quite a lot more to say about the matter, and which we discuss in Chap. 4.

1.3 The Rich Diversity of Lived Experience

The potential diversity of lived experience as personally felt is, by definition, vast. The interviews with the British flood victims demonstrate shared experience of the event and its aftermath, but in the detail they are all different in how they were affected by it and how they have, and still are, coping with it. This is obvious because no two people are the same, but we need to go beyond the obvious. If it is going to further our understanding of climate change, and perhaps also inform policy on it as later chapters of this book argue, we need to explore patterns of lived experience at a social as much as at an individual level. The four case studies that follow concern the fortunes of a poor person in the United States who was caught up in Hurricane Katrina, an affluent community in the United Kingdom protesting against a wind farm proposal, two vulnerable forest dwelling communities in East Africa and two African environmental activists.

1.3.1 Contrasting Lived Experiences of Rich and Poor

One common pattern to which the reader-blog comments associated with The Guardian newspaper interviews pointed was that the flood victims appeared to be affluent. Some might suffer long-lasting psychological scars from the event, but all appeared able to pull through materially. This raises the question of what the contrasting lived experiences of poor people might be in relation to an extreme weather event.

August 2005, Hurricane Katrina hits New Orleans in the southern United States

Over eight years prior to the United Kingdom 2014 floods, Hurricane Katrina swept over the Gulf of Mexico and the far south of the United States. Over 1,800 people died and the property damage alone was more than 100 billion dollars.

The city and port of New Orleans received a direct hit from Hurricane Katrina. The levee—the embankment protection against flooding from the River Mississippi—failed, resulting in substantial flooding. Many thousands were forced to evacuate, and, as services broke down in the immediate aftermath, reports emerged of looting by those who remained.

Prior to Katrina, New Orleans was characterised as a poor city on a variety of quantitative indicators. Its population was declining (Liu and Plyer 2008: 11), job opportunities were weak (Muro et al. 2005), the average weekly wage in the private sector was 11.6 % lower than the comparable national average wage (Dolfman et al. 2007: 5), public transport outside of the tourist areas was described as inadequate and education as 'one of the nation's worst' (Plaisance 2006).

The above paragraph describes briefly the economic and social context in which Iversen and Armstrong (2009) examined the 'lived experiences' of three citizens prior to Katrina and during its aftermath. Joseph, a 23-year old High School graduate from an immigrant background and a married father with two young children, was one of the three. Here is a highly edited account of his story as told by Iversen and Armstrong (Ibid.):

> After a bad experience on an official construction apprenticeship scheme, Joseph found a construction job through the church to which he belonged, which acted to facilitate both emotional and practical networks among its largely poor congregation. His new boss was a trusted church colleague and self-employed contractor, who also rented Joseph and his family a house that he owned
>
> Come Katrina, the house was completely demolished by its force. Joseph and his family evacuated. Nine months later it appears that he had not returned to New Orleans (and Iversen and Armstrong seemed to have lost contact with him). In any case, as further reported by the authors, residential construction work in post-Katrina New Orleans was rare, and what existed was dominated by people who had recently moved to the city from elsewhere in the United States.

Comparison of Joseph's story with those of the UK flood victims of 2014 illustrates graphically how much economic, social and cultural context influences the lived experiences of those who are caught up in these events. The UK stories do, indeed, seem to be of affluent people. The photographs accompanying the interviews show large houses and gardens, big cars and so forth. There can be no doubt that they were severely disrupted, but they have the resources to rebuild, at least materially, their lives. Even a business that was physically wrecked has been rebuilt. Some even report positive outcomes, for example a greater sense of community.

The Joseph-Hurricane Katrina narrative by contrast shows a vulnerable person for whom rebuilding is much more of a challenge. This is not to say that he and those who have similar circumstances are necessarily passive as Chap. 5 will show. It does illustrate in Joseph's case that, even if you live in the richest country on the planet, insecure livelihoods, weak education and poor housing combine to make greater the negative impacts of extreme weather events.

While these examples tell us something about the drivers of lived experience, the reverse is also true—the lived experiences tell us something about extreme, climate-related, weather events. That is, an extreme weather event is not simply what Quarantelli (2005) describes as a 'focused occasion in terms of time and space', for its impacts are embedded in societal structures, where the already vulnerable are the most vulnerable to the event.

1.3.2 The Lived Experience of Public Attempts at Climate Change Mitigation and Adaptation in an Affluent Country

It is common to think, as in the previous examples, that humans experience climate change through its direct impacts. While this may be true for many, for others the most tangible, directly felt experience is not of a flood, a drought or a huge sea wave, but of attempts to mitigate[1] through coordinated action the perceived primary source—anthropogenic (i.e. caused by human beings) global warming—of such events.

Since the early 1990s, governments have met regularly under the auspices of the United Nations to try and reach international agreements for action on climate change. Sometimes they get somewhere, at other times they seem to go backwards. Usually, the issue concerns a target for limiting the emissions of the main global warming gas, carbon dioxide, so that the earth's temperature does not increase beyond a certain magnitude. For some years, this target has been set in international forums at 2 °C. This is considered to be the allowable temperature rise that would avoid so-called 'dangerous global warming', where anything more

[1]As well as mitigation, which usually refers to reducing carbon emissions so as to limit global warming and its consequences for the climate, human action also involves adaptation to current and potential future impacts. Examples of the latter include flood defences, breeding drought-resistant crops and so forth, Adaptation and mitigation are considered further in Chap. 6.

would lead to massive societal disruption. Needless to say, it is a contested target, but given its prominence in international policy making on climate change, let us for our purposes here accept it and examine its ramifications in relation to mitigation action.

Various mechanisms are proposed for meeting the 2 °C target. For example:

- Establish an agreed cost of carbon (shorthand for carbon dioxide emissions) that is applied internationally and which would act as an incentive for innovation to reduce emissions. As British economist, Nicholas Stern put it, if public policy gives the right signals and rewards for cutting greenhouse gases, 'markets and entrepreneurship will drive the response' (Stern 2010: 99).
- Tax carbon-emitting fuels.
- Provide incentives for change in consumer habits towards 'carbon-neutral' goods and services.
- Invest in 'green' renewable energy infrastructure that has low or no carbon emissions.

It is not the purpose of this book to enter the specific debates about these mitigation options which are well documented elsewhere. For a good economic analysis, see Stern (2010). For a full scientific analysis, see the reports that are published periodically by the IPCC, the latest being in 2013/2014.

Instead we illustrate how various measures that are designed to reduce carbon emissions interact with people's lived experiences of them. As we write, there is still no internationally agreed and binding protocol for putting a price on carbon. Although, if eventually agreed, it will filter into citizen lived experiences, we do not dwell on it here. Similarly, a debate rages over so-called green taxes on energy. In that they raise the cost of energy, they do of course act as an incentive for consumers to conserve at home and drive fuel-efficient cars. They are also intended to help pay for the necessary infrastructure change. They are, however, politically unpopular in those countries that have adopted some form of 'green' energy taxation.

The most direct and obvious interaction with lived experiences comes when new energy infrastructure and other public measures to mitigate carbon dioxide emissions appear in our back yards, or near-back yards. These experiences frequently appear as negative and generate significant opposition to the proposed measures. They have spawned a field of academic study on social acceptance of renewable energy, particularly in relation to clusters of wind turbines (wind farms) where opposition seems to have been most entrenched in the United Kingdom. For a review of the complexities involved in social acceptance, see Fast (2013). While accepting his point that people play other roles in relation to renewable energy projects—for example as investors, producers and users—this does not alter the fact that a significant and vociferous role is oppositional, and that these voices will continue to be heard.

Here is one example from the Tove Valley in the Midlands of England, where a 'Tove Action Group' of local residents in 2014 declared a 'victory' on their website (Tove Action Group 2014) after the developer decided after three years not to proceed with a planning application for the development of a wind energy farm. The Action Group campaigned with 12 'reasons to say no':

1. A negative impact on the local landscape, which the Action Group describes as having 'historical and cultural importance as well as including woodland, natural features and farming'
2. Intrusive visual dominance of eight large turbines that were proposed
3. Distance impact, as the turbines might be as close as 500 m from the nearest residential houses
4. Devaluation of property, especially residential houses that would lie close to the development
5. Invasive noise that would be generated by the turbines
6. Disruption during construction
7. Birds and bats being caught in, and killed by, the turbine blades
8. Potential negative impact on local quality of life and well-being
9. The possibility of intrusive 'shadow flicker' caused by sunshine being reflected off turbine blades. This effect could be further associated with health problems
10. The perceived unfairness of providing government subsidies to wind farm developers
11. Perceived misinformation regarding positive efficiency claims for wind power
12. Wind speeds in the area have been below the national average, making this an unsuitable site.

This is quite a mixture of 'reasons to oppose'. The first seven concern probable direct impacts on lived experience, the eighth might result from a combination of these. The ninth concerns reports from wind turbines elsewhere, and 10–12 make claims about wind farms in general.

Again, we must stress, that it is not our intention to provide an analysis of each of these 12 'reasons to oppose' but to note that they appear to link significantly to a lived experience view. As with all infrastructure projects, which include airport expansion, new fast roads and high-speed train lines among others, 'green' energy courts controversy where opposition usually starts with those whose lived experience will be directly impacted by the development.

The Tove Valley of England is an area of relative affluence in a relatively rich country. Mitigation measures may also appear in some of the poorest countries of the world, incentivised by international development initiatives. Yet again, these schemes are often controversial, and impact on lived experiences. The following case study from Kenya and Uganda is taken from the doctoral dissertation of Nyukuri (2013).

1.3.3 Connecting Climate Change and Climate Change Mitigation to Lived Experience of Vulnerability of Tropical Forest Communities in East Africa

This case study concerns how international climate change mitigation schemes that involve combating the loss of tropical rainforests impact on two poor communities, the Ogiek in Kenya and the Batwa in Uganda. The two communities have strong historical ties to the forests as sources of livelihoods.

International and national climate change contexts

Both Kenya and Uganda are described as poor, although by no means the poorest, in terms of world development data. Thus, the Human Development Index of the United Nations Development Programme, a composite of income, health and education, ranks Kenya at 147 and Uganda at 164 out of a total of 187 countries (UNDP 2014).

Undoubtedly as a result of relatively low levels of economic activity and relatively low population densities, Africa's greenhouse gas emissions are small compared to those of affluent Northern Europe, North America, Australasia, Japan and, in recent years, China (Toulmin 2009: 7–8; IPCC 2007). Greenhouse gas emissions know no borders, however, and the continent is already suffering impacts of climate variability, mainly because of its high dependence on agriculture and other natural resources that are sensitive to climate—land, water, forest, wildlife, biodiversity and energy (Boko et al. 2007). Kenya and Uganda are part of East Africa which in recent years has recorded an increase in droughts, floods, windstorms and mudslides (IPCC 2007).

Small though the greenhouse gas emissions of each of these countries are, they are still significant. In Uganda and Kenya, as elsewhere on the African continent, deforestation (Box 1.1) is one identified source. Tropical forests act as a carbon store, or 'sink', through their absorption of atmospheric carbon dioxide during the process of photosynthesis that drives plant growth. This 'carbon sink' is obviously lost when forests are cleared during human activities. Worse the clearing process often involves burning or leaving the trees to rot, which results in releasing their stored carbon dioxide back into the atmosphere.

This is a real problem. The United Nations Food and Agriculture Organisation (FAO 2011) estimates that between 1990 and 2005 Africa lost over 9 % of its forest cover. The IPCC report of 2007 estimated that nearly 20 % of global anthropogenic carbon emissions result from the forestry sector which includes deforestation (IPCC 2007).

It is no surprise, therefore, that schemes to prevent further deforestation and to reverse the process through financial incentives to plant trees have come to the fore in many areas of the world, including Kenya and Uganda. The stated aim is rightly to contribute to both reducing and offsetting carbon emissions. Gregersen et al. (2010) also suggest that it is popular because it is a relatively low-cost option.

One such internationally supported scheme is the United Nations 'Reducing Emissions from Deforestation in Developing Countries' (REDD) which was first promoted in 2005, and its enhanced version REDD+[2] since 2007. The way in which REDD works varies from context to context but, in Uganda, Mwayafu and

[2]There is some confusion over terminology. REDD+ is also known popularly as Reducing Emissions from Deforestation and Forest Degradation (plus some further supplementary text). For the purposes of this book, there is no need to enter into this issue.

Peskett (2009) report that, through an intermediary non-governmental organisation (NGO), individual land owners enter a contract with companies from affluent countries to sell them 'carbon credits' through an afforestation and reforestation (Box 1.1) programme. In this way, the landowners are recompensed for afforestation and reforestation, while the companies concerned are able to use the credits that they buy to offset the carbon emissions from their regular activities, thus contributing to meeting targets in their home countries.

Box 1.1 Deforestation reforestation and afforestation

Deforestation is simply loss of forest cover, usually connected to human activities, for example commercial logging. *Reforestation* is replenishing forest that has been lost. *Afforestation* is converting land not previously known to have been forested or, if it has been it was a long time ago, into forest.

*Nyukuri's (*2013*) case study of forest communities in East Africa*

Historically, the Batwa of Uganda inhabited the forests of the south-west of the country. The majority of the Ogiek in Kenya live around the Mau forest to the North-west of the capital Nairobi. Both groups were originally hunter-gatherers (Blackburn 1976).

Drawing on Nyukuri's dissertation, the lived experience of the Batwa has been influenced by the following elements (numbers in brackets represent page number references to the dissertation):

- Historical invasion and displacement by neighbouring groups from the forests that they inhabited. This goes back to pre-colonial times. During the colonial period, Belgium had a presence before Uganda became a British Protectorate in 1894 and there are examples where Belgium supported encroachment of Batwa lands. During their rule, the British authorities established protected areas in the forest, ignoring any rights of the Batwa to the land.
- Uganda gained independence in 1962, but by this time the Batwa were largely 'landless, homeless, economically weak and dependent on their former enemies [that is, stronger, neighbouring groups] for survival' (130).
- In 1991, the Ugandan Government created three national parks in the area for conservation purposes. The Batwa were forcefully evicted by law. They had no formal property rights to the land and were not compensated. Citing a study by Kabananukye and Wily (1996), Nyukuri reports that 'amongst all the Batwa who were displaced from the three parks, 82 % were entirely landless. The majority of these (80 %) lived on the land of other farmers, 9.4 % lived on government land and 10 % on church land (133). This meant that the majority were essentially squatting, where they 'had to rely upon the patronage of their farming neighbours to provide land for them to live on' (133). Thus, she quotes:

> *Government did nothing but only evicted us from what belonged to us* (Batwa middle aged male A).
>
> *There is no land for us to get for ourselves* (Batwa middle aged male B).
>
> *The government said the forest belongs to the animals and we had to leave* (Batwa elderly male).

- Later (2001–present) management plans for the National Parks have attempted to integrate the Batwa, including access for certain products (herbs and honey—their traditional livelihood sources), but also to exploit their local knowledge 'in the form of employment to provide services such as tracking and guiding' (135). At the time of publishing the dissertation, however, there were, no Batwa employed in the park—a result of low levels of education, particularly illiteracy, and no spoken English. Some Batwa were entertaining tourists as dancing and singing groups.
- Current climate change mitigation policies in the forests that involve the United Nations REDD+ scheme have further entrenched the legitimacy of displacement or restricted access, while rewarding those with formal ownership of the land. It is no surprise, therefore, that campaigns by the Batwa, who have no land titles, to be allowed to return to the forests and access their products have remained largely un-met. Nyukuri concludes, 'Under these circumstances, the Batwa are expected to adapt to the livelihoods practised since eviction such as farming and to exert their agency in new environments' (137).
- With farming, however, there is the further climate change dimension of increased weather variability, including floods and prolonged droughts. Thus, Nyukuri notes, 'In addition, the Batwa are expected to diversify the types of crops they rely on and plant drought-resistant crops, observe seasonal changes and plant crops according to the changing seasons' (137). Lack of land titles, however, again puts them at a disadvantage. If they are mainly squatting on land owned by other farmers, they may be evicted at any time and there are no incentives to invest in the know-how that is required. This extends to the houses that are erected which are unable to withstand the mudslides that are common during prolonged extreme rains:

> *It sometimes causes bad floods, which then destroy our crops just like last year* (Batwa middle aged male A).
>
> *In the past there was enough rain. When it rained the fields would yield all kinds of fruits. But now things are different. Cows are dying. The rains have disappeared* (Batwa middle aged male B).

- The situation is bound with, and compounded, by other factors. Discrimination and segregation from neighbouring communities is rife, with a strong sense that the Batwa are primitive and non-human. The Batwa have to collect water from different sources (Turyatunga 2010), while eating, drinking and inter-marrying with non-Batwa is out of the question (Kenrick and Lewis 2001). In education, the 2002 census reported that 40.4 % of Batwa in Uganda had never been to school. Of those who do attend, a large proportion abandon schools as a result of discrimination and bullying (Tumushabe and Musiime 2006; Kabananukye 2011; both cited in Nyukuri: 147–138). In health, Nyukuri cites a 2006 report from the British medical journal, *The Lancet*, which found that up to one in four Batwa children dies at birth in Uganda, nearly four times as frequently as the rate for Uganda as a whole (Ohenjo et al. 2006). Overall the health of the Batwa is among the poorest in Uganda.

- Many Batwa have internalised the external perception of their inferiority. Evicted from their historical lands, squatting in makeshift houses, poorly educated, in poor health and feeling incapable of adapting to their circumstances, they simply beg as a way of life. For example:

> *Forest life was better because we had enough food and land. Nowadays we only depend on begging and sometimes those we beg from refuse to give us* (Batwa elderly female A).

> *In the past we would get food from the forest but now we are just roaming, nowhere to go and nothing to eat* (Batwa elderly female B).

> *The entire community of Batwa are poor and depend on begging as a form of livelihood. Their housing conditions are very poor and lack access to social services such a running water and health facilities. Their children experience high rates of malnutrition and there is societal discrimination against them* (Batwa key informant from a non-governmental organization).

In the language of climate change policy documents, the Batwa have weak capacity to adapt to changing weather and to the opportunities provided by mitigation activities. The above summary of their lived experiences tells the story of why. Climate change is not the cause of their vulnerability, but certainly exacerbates it.

Turning now to the lived experience of the Ogiek in the Mau forest of Kenya:

- Recorded instances of the Ogiek being deprived of their historic lands go back to 1856 when neighbouring pastoralists, the Maasi, annexed part of the Mau forest for their own exploitation. The British formally arrived in 1895 and ruled until 1964. The colonial authorities presided over three waves of evictions between 1911 and 1927, declaring the forests as (British) crown lands. During this period, indigenous trees were felled, often being replaced by exotic varieties that were more suited to cutting for timber. Some pockets were designated by the British as natural reserves and protected (Sung 2004).
- On eviction, the Ogiek became assimilated into neighbouring communities. According to Nyukuri (148), this became formal colonial policy after the Kenya Land Commission recommended in 1933 the relocation of the Ogiek to reserves of bigger ethnic groups with whom they had close affinity (Government of Kenya 1933).
- Whether by deliberate design or not, the assimilation into larger ethnic groups, created a situation after independence whereby the Ogiek were unable to make claims as a group to resettle in their traditional habitat.
- Since independence, the Mau forest has been controlled by many groups intent on exploiting its resources, including timber companies. In parallel, since 1991 the Kenya government has resettled groups on land as the forest has been cleared. These groups have included the Ogiek who have been given small

parcels of land to farm, but also many other groups. Often, allocation has been politically motivated, leading to clashes between groups. Nyukuri (150, 151) states that there has been 'constant conflict with the Ogiek who see the destruction of forests and the alienation of their lands as a continued threat to their existence'.

- While they have been resettled by the government on cleared forest land and farm it, the Okiek do not have formal titles of ownership, which makes them susceptible to changing government policy.
- Thus, in relation to climate change mitigation measures that involve REDD+ schemes in the area, thousands of Ogiek have been forced to abandon their homes in the Mau forest because of conservation measures (Ochieng 2009; Collins 2009). The issue is further complicated because the Mau region as a whole is a nationally important water catchment area, and the government is seeking to manage it more actively in the light of changing weather patterns. Plans to restore the forest to its original position to curb further degradation of the Mau ecosystem do not, however, present opportunities for the Ogiek to settle in their original homeland (Kameri-Mbote and Nyukuri 2013). Nyukuri concludes: 'Climate change and climate change policies are therefore exacerbating their already precarious position' (154).

The lived experiences of the Batwa and Ogiek have some elements in common with Joseph from New Orleans. In each of these examples of lived experience, historical and existing vulnerability appears as the prime challenge when faced with the extremes of climate variability. In the example of Joseph, this seems to be the case—and note that he has no formal ownership of property and land either—in the richest country in the world.

In contrast, the Batwa and Ogiek experiences appear to bear no resemblance to those of the eight United Kingdom flood victims of 2014. These UK citizens are all relatively affluent, even in relation to British households nationally. They appear educated. They formally own their properties. They have government help and all except one has property insurance, even if completing the claim forms appears to be a bureaucratic nightmare. None of them, except one, farms and hence are not linked directly to a climate-sensitive livelihood. If there are any long-term scars, they are emotional. To them the floods constituted a rare traumatic event. To the Batwa and Ogiek, droughts and floods simply exacerbate existing and historical vulnerability. Interestingly, Nyukuri observes that, during her study, the Batwa respondents rarely pointed to the impacts of climate change but to other problems such as poor living standards and lack of food. For the Ogiek, conflicts over land, how it was shared and used with neighbouring communities were of particular concern rather than climate change.

A further strong contrast lies in the UK experiences and Batwa and Ogiek experiences of climate change mitigation programmes. Each of these in this chapter has presented a negative experience. To the UK opponents of the wind energy farm, however, the threat is principally to loss of amenity affecting adversely quality of life. To the Ogiek and Batwa, the threat is to further alienation from the forests with which they have strong historical and cultural ties, and which was historically a direct source of their livelihoods. The UK opponents, moreover, opposed successfully in advance the imminent threat of the wind farm, whereas for the Batwa and Ogiek, forest conservation and schemes have already had a negative impact on them.

1.3.4 The Lived Experience of Environmental Activists

Many individuals become active agents for the broader changes in society, business and politics that they argue are needed for meaningful impact on climate change and other environmental issues. These are the environmental activists who mobilise for effective action, often through campaigning NGOs.

Below, two environmental activists from Sub-Saharan Africa document the experiences that made them what they are today. We follow their accounts with our comparison:

Charlene Hewat's account (Zimbabwe)

I am a White African, born in 1963 on a farm in Zimbabwe, which when I started out had nothing on it except for the flora and fauna. To begin I lived in a mud type house and my bath was a tin tub, which we had to heat on the log fire. We had no running water or electricity and in fact most of my childhood was spent with a simple generator for power. Our food we would source mainly from the wild and grow our own vegetables and eventually fruits. Coming from a farming background I remember the importance of the rains and how we would celebrate the first rains, no matter where we were, by getting ourselves soaked. We carried out dry land farming, so rains were very important to us and still are today. Every day we would measure the rain gauge and keep a record of the rainfall annually. In a nutshell we depended a lot on our natural environment for our survival.

My father was a born environmental, earthy type of person and was a real inspiration to me. When we started farming he would always talk about how everything was interconnected and how we needed to farm in a way that did not harm nature and the environment. Papa, as I called him, would leave 5–6 m of natural bush between each of the fields, he would not let anyone bulldoze the anthills and left them standing in the middle of the fields. He would explain how important nature is to the production of crops, the bees to pollinate, and the trees to absorb the heat and cool the earth. He even refused to plough the lands and at the time carried out what was called minimal tillage. When I look back I realise that all that he was teaching me was about how to conserve and take care of our environment and today the talk amongst agriculturalists is about zero tillage and how it can help towards carbon sequestration.

My passion for the environment and the love of nature was instilled in me at a very young age, thanks to Papa, but he also taught me about the importance of people and the planet. How people were destroying the planet and yet at the same time how the traditional leaders were real protectors of the planet. He told me about the story of the local chief not allowing anyone to cut trees without permission and if they did they were punished by sitting on top of an open anthill. Yet today I see throughout our beautiful country thousands of trees being chopped, some out of necessity but most of it out of greed. Something that is still vivid in my memory is when I accompanied Papa, aged 14 years to visit the local chief, Chief Chirau, and there before us in the maize field was the Chief in a brand new Mercedes with a plough attached to it, ploughing his fields. I did a double take, was this for real? What have we done to our people? I said. What was this all about—technology advancing at a faster rate and us not being aware or capable of keeping up with it?

When I look back at my younger days, I feel privileged to have been born in a world that was natural, raw and real. Visiting many wildlife areas, it was the one thing I loved and in my early 20s I came across some poached rhino. Horns hacked off and the carcass left for dead. This triggered something inside of me and I went to find out from National Parks what was being done about it. Not much as there was little finance and equipment to cope with the high volumes of poaching. It was here that I decided to do something about it and rode a bicycle from the United Kingdom to Zimbabwe, 22,000 km, with a colleague of mine. We raised 250,000 British pounds for rhino conservation but for me what was amazing was the lessons along the way. I noticed how Europe was busy, with polluted cities, and people had no time for each other. There were all the latest up-to-date gadgets around, and very little wildlife and open spaces, mainly concrete. Whilst travelling through Africa there were vast open spaces, wildlife and friendly open people who would welcome us into their homes, and although they were poor they would provide us with food and water. I remember on the one day we asked for water, they said just over the hill, well no water was seen for miles and miles and eventually we had to dig for water in the sand bed. The differences between the developed and the developing countries was hugely noticeable and although more consumerism, pollution is happening in the developed world, it seems that Southern Africa is going to be the most affected with regards to climate change.

Currently I am working with a local Environment based organisation and this is when I first got to hear about climate change some 15 years ago. Everyone was making such a big issue out of climate change and the negative impacts it is having and going to have especially in Africa. For me, I think climate change is a natural cycle that the world is heating up, however, what we as humans have done and are doing is speeding up the process. Our greed, consumptive patterns and basic loss of connection to our planet are all contributing to climate change.

There is no doubt that climate change is real, especially here in Africa. I have noticed the rainfall patterns changing, the seasons changing and geographical regions changing. With the work that I do, I travel extensively out into the rural community areas and meet with our many community farmers who have their stories to tell about climate change. In the Eastern Highlands, the rural farmers are telling us stories about how they are now having two rainy seasons and not one and that there is less and less rain falling. In Hwange, Matabeleland North, the farmers talk about changes in the vegetation and the increase in temperature.

It was about 12 years ago when we as an organisation decided to start introducing, and in most areas reintroducing, small grains to the farmers. I noticed back then the reduction in rainfall and wanted to look at how farmers could adapt to the changing climate. Small grains require little rain and. if farmed with zero tillage and conservation agricultural methods, the likelihood of reaping a crop is higher than if one used the normal hybrid maize, tilled soils and fertilisers. It was difficult at first to convince the farmers that this is the way forward, so we started with a few farmers and with their success more and more farmers took up small grains and conservation agriculture methods.

Climate change adaptation is now real and something all development and development agencies need to incorporate as part of their interventions. It is not a singular issue, a separate issue, but an inclusive one. International organisations often come into Africa with quick fixes, immediate short-term solutions which in the long term not only fail but also break up and destroy communities. They think they are doing the right thing at the time. A good example is food aid. People are starving so let's go out and give them food instead of giving them small amounts of food for a specific time frame whilst at the same time teaching them how to feed themselves. The international organisation feels good, it has fed some targeted hungry communities, it has reported back to the donor country and it has fulfilled what it thinks is real development. The question is how donors are going to deal with climate change issues. Recently when we received from a donor seed to distribute we received hybrid maize seed, as that is what the donor was giving out, although we specifically requested small grains. What do we do in this instance? Basically the big brother rules, but who suffers in the long term? The community.

Adaptation is critical in Africa and over 5 years now I have seen an increased uptake of small grains by local communities. The communities are realising that climate change is real and to deal with it they are going to have to adapt. Small grains is one of the adaptations that is taking place more and more in the areas where we are now working and working well. The communities who are undertaking small grains and conservation farming methods are surviving, whilst others are producing very little or no crops with hybrid maize seed. Interestingly, the seed companies are now looking more and more at small grains and at how they can get involved in the production and distribution of small grain seed.

One big area, which is lacking with regards to climate change, is biodiversity monitoring and banking. We, Environment Africa (the NGO where Charlene works), as an organisation are looking at establishing a bio bank so as to bank biodiversity for the future. The project is in the initial stages of development and new technologies of banking biodiversity at room temperature are currently being explored together with the monitoring systems of climate change along the Zambezi River. This project will not only enable us to bank biodiversity for the future but will also enable us to work towards climate change adaptation as we will know what it is we are dealing with. This is an exciting project for the future.

Climate change must not be driven as a single issue; it is too big and too daunting to us all. What we need to look at is how to integrate climate change into our daily lives both at work and at home.

Richard Kimbowa's account (Uganda)

I was born and grew up in the central part of Uganda 40 years ago. From a background of a small scale agro-based family that depended on a combination of crop and animal husbandry for a livelihood in Wakiso district near Uganda's Capital of Kampala, I was able to study until university level.

During my high school I used to attend some school geography club events that included monthly debates on a wide range of topics. In some way climate change discussions featured. More frequently, during my undergraduate studies where I studied forestry, I was able to learn a little more about climate change issues from this professional standpoint.

This exposure enabled me to attend several discussions, trips organised by nature related clubs at University and numerous talks and discussions that helped me understand the topic more, although available literature then was limited to experiences from other countries (especially Asia and Western Europe).

All my family's agricultural work has been dependent on the two rainy seasons which fall around April and November, combined with the relatively fertile soils that have until recently supported a satisfactory crop yield and animal production. However, changes in the intensity and predictability of the rainy season in this largely rain-fed agriculture have severely interfered with family incomes and hence affected livelihoods (including dependants who are far away in the rural areas).

Rainfall failure and its increasing irregularity as evidenced by sporadic occurrence, strong winds and hailstorms is a major climate factor of interest to me, coming from an agro-based background. Climate change related impacts have exacerbated the decline in agricultural productivity. This takes the form of failure of the sown crop (grain) seed to germinate, poor harvest due to excess rains/ extended drought conditions, poor post-harvest handling leading to huge losses and/or poor prices from the crop produce. In relation to animals, the water stress arising from extended droughts meant that poor quality water had to be looked for from far distances and at awkward hours (for example staying at a water point up to late in the night or very early morning; and having to come back early from school in order to be able to look for water), which affected my concentration at school. In some cases the water collected was of questionable quality, but we had no alternative to use it for cooking, bathing, washing and other domestic purposes.

Effects of climate change have been compounded by the rolling back of the Ugandan state in the 1990s (no agricultural extension service staff, no more farm inputs, etc.) with dire consequences to small scale farmers like my family, who have to turn to private advisers at a cost.

Given my background and the experiences of the changing climate (especially rainfall patterns on which many Ugandans depend for their agricultural activities), I feel obliged to work with like-minded individuals and institutions to influence national plans that address adaptive capacity of the most vulnerable communities (small scale farmers, fishers, rural based societies and the urban poor). Fortunately, I work for an NGO—Uganda Coalition for Sustainable Development—where such a chance exists through national, regional and global campaign efforts (provision of information on available coping options, research, awareness raising, sign-ups, among others).

It is clear that this is going to be part of our life. What is therefore important is to have in place measures and options that can enable citizens to cope with this new development challenge at their own level (agriculture and food security, water supply and sanitation, transport, health, and so on).

This is not an easy thing to have in place in developing countries like Uganda where available safety nets (sometimes even to levels of disaster preparedness) lie with families that are already stretched by declining food yields, disease outbreaks, rapid urbanisation rates and the rolling back of the state in the much needed social sectors like agricultural extension and healthcare. In this regard, I feel that there is a global responsibility from the rich countries and well-off citizens to help offset this need.

A comparison of the two accounts

The contemporary lived experiences of both Charlene and Richard are heavily influenced by their family histories, especially their rural childhoods where their survival was connected directly to their relationship with the biophysical environment around them. As a result they both appear to have a profound respect for the biophysical environment. Charlene mentions specifically the role of her father in teaching and showing her the interconnectedness of nature, while Richard obviously learned about these things also through formal education at high school and university.

Charlene's pride that the family farm was built from nothing probably instilled in her a profound sense of active agency to bring about change. Thus, as a young adult she campaigned to save the rhino and since has been prominent in an environmental NGO. Richard concentrates on the difficult material circumstances of life in rural Uganda, the withdrawal of state services (such as extension workers) and the sense of loss of control due to climate variability. These have made farming communities more vulnerable but he too has decided to contribute towards solutions in later life.

Charlene's account suggests that numerous triggers at key moments in her life turned her into an activist: her father's insistence of farming with nature rather than against it; the visit to the local chief who was ploughing his field with the aid of a brand new Mercedes car, the misguided interventions of international organisations, and above all the sight of the poached rhino. Regarding climate change specifically, she obviously soon absorbed the messages about it from the international media and elsewhere, but interpreted them in her own way. In contrast, Richard does not mention specific triggers, but the general conditions of hardship in rural areas which have been exacerbated by climate variability. Together with his formal education, he has felt obliged to work with like-minded individuals who wish to address the situation. He also works, therefore, for an environmental NGO.

While the influences, broader and direct, on both Charlene and Richard have been profound, their experience is obviously evolving through action-learning cycles. Charlene refers specifically to the development of drought-resistant small grains and also for the need to help everyone to adapt for themselves. Likewise Richard is explicitly concerned to enable citizens to cope, and interestingly he conceptualises climate variability within the broader development challenge that all poor countries face. However, while Charlene is critical of the work of many international development aid agencies in Zimbabwe, Richard feels that the international community has a responsibility to help. It is possible that this reflects a deeper influence on their experience, between Charlene's sense of active agency, and a sense in Richard of injustice towards his country for which the international community bears significant historical responsibility.

1.4 From Flood Victims in the United Kingdom to Small Farmers and Drought-Resistant Food Grains in Zimbabwe: Making Sense of the Lived Experience of Climate Change

We all have lived experiences. To some extent they are highly personal, but to some extent they are also shared with others of similar circumstance. How can we stand back and make generalisations in order to understand them better, and also to make them useful as a form of knowledge which policy makers in Government, for example, might be able to make use of? At first glance, the examples and case studies in this chapter illustrate only how varied they are, and by consequence what a minefield making sense of them will be.

Our starting point is to look for patterns in the accounts, even if they appear to be somewhat abstract. It is from these patterns, however, that we may start to make generalisations about lived experience which eventually will enable us to conceptualise it and put it to use. Below, we describe seven that emerge from the accounts, followed by their synthesis into a working conceptualisation into the lived experience of climate change:

Pattern 1: the influence of context
All of the accounts suggest to greater or lesser extents the important influence of social and geographical context on lived experience. It is bound up with where we live in the world, our degree of poverty/affluence and hence our vulnerability/resilience, our family circumstances and the lifestyles expected of us, our degree of dependence on natural resources for livelihoods, and our capacity to adapt.

Pattern 2: the influence of personal histories
Contemporary lived experiences of climate change are not simply the experience of particular events, but evolve from previous histories. This is explicitly true for

those who become activists, as with Charlene and Richard; for those with histories of loss such as the displaced forest communities of Kenya and Uganda; and for those with other histories of vulnerability, for example the story of Joseph in relation to Hurricane Katrina in the United States. The histories of the affluent victims of the United Kingdom floods or those who opposed the wind farm are not expressed explicitly, but it seems safe to assume that they too have influenced their lived experiences.

Pattern 3: the lived experience of climate change cannot be isolated from other lived experiences—it is relational to them
This is most directly expressed in Elvin Nyukuri's account of the forest communities in Kenya and Uganda, where climate change, and specifically actions to mitigate climate change, reinforce a more general experience of vulnerability. Similarly, climate change in Charlene Hewat's and Richard Kimbowa's accounts is intertwined with the overall experience of farming in Sub-Saharan Africa. Joseph in New Orleans was also already vulnerable before Hurricane Katrina hit.

For the accounts that are related to affluence, we might note that the lived experience of climate change is likewise interwoven with other experiences. Thus, the interviewees of the Guardian newspaper similarly express their experiences of the United Kingdom floods in terms of their businesses, their dream homes and their ways of life.

We always have to consider the lived experience of climate change as relational to these other experiences rather than as something we can distil and discuss in pure form.

Pattern 4: lived experience in relation to climate change as a battle
To some extent this pattern is present in all of the accounts we have seen in this chapter. It is most explicitly evident, however in the accounts of our two Sub-Saharan African activists, Charlene and Richard, where beleaguered rural communities are being forced to adapt. It is also evident in the opposition to the wind farm, where the community feels beleaguered in a different way—it is being asked to sacrifice its lifestyle. This idea of climate change representing a battle, or even a war, that inevitably leads to the idea of sacrifice in order to fight it is quite dominant in public commentaries and books on the subject. We call it a dominant way in which climate change is 'framed', and which in turn influences how we think about it. We explore the battle/war framing further in Chaps. 6 and 7

Pattern 5: lived experience is collective as well as personal
Despite their differing accounts in the detail, broadly speaking the United Kingdom flood victims had a shared experience. In two cases this even resulted in positive outcomes—a sense of community was revived, as was a collective spirit to ensure that a business was not lost. The other accounts also illustrate the collective nature of lived experiences—for example the forest communities in Kenya and Uganda, the community opposition to the wind farm in the United Kingdom, and the shared experiences of farmers in Uganda and Zimbabwe.

Pattern 6: the influence of engagement with others and reflection
Pattern 5 on shared experience implies that we share, that is, we communicate and engage with others. This is true of all the examples given for pattern 5. Moreover, engagement means that we reflect. We observe reflection in the Guardian newspaper interviews of British flood victims and the quotations that accompany the stories of the forest communities in East Africa. Our ability to reflect and engage with others is what the philosopher Jürgen Habermas, among others, claims make us human. It also implies that lived experience is dynamic. We change through reflection and engagement with others.

Pattern 7: the influence of action
All of the accounts refer to action in one form or another and it is clear that lived experience is linked significantly to such action. It is useful to distinguish two kinds of action from the accounts:

a) Reactive action which is in response to climate-related impacts. For example: the re-building activities of the United Kingdom flood victims; and the opposition to the wind farm proposal, again in the United Kingdom.
b) Proactive action to mitigate climate change or to put adaptive measures in place before the impacts are felt: for example, the work of Environment Africa in Zimbabwe.

Synthesising a working conceptualisation of climate change
We cannot change our histories (Pattern 2). It is also very difficult to change, at least in the short-term, our social contexts (Pattern 1), the dominant way in which climate change is framed (Pattern 4), and the ways in which the lived experience of climate change is interwoven with other lived experiences (Pattern 3), some of which may be considered more pressing—for example the everyday experiences of poverty and vulnerability. If these four patterns constituted the only, or even the main, drivers of the lived experience of climate change, it might become wholly predictable and relatively static. We could feed into a computer our social contexts, our personal histories, the other pressing issues we face in everyday life and the dominant framing of climate change, press 'enter' and all would be revealed.

Life, however, is not like that. We are also influenced directly by the experience of immediately-felt events, such as sudden extreme weather, the severity of which cannot be predicted far in advance. Moreover, as human beings we share, reflect, communicate and engage with others (Patterns 5 and 6) and we take action (Pattern 7). The action we take then leads to further reflection and so the process goes round in a circle. It is the basis of an experiential action-learning cycle as articulated by Kolb (1984), although the rationale as we go through the process is not necessarily evident to anyone but ourselves.

This makes our lived experience of climate change unpredictable and interesting. It also means that our lived experience is dynamic. The influence of context, history, other pressing challenges and dominant framing will always be there. So too will unpredictable weather and impacts, but we can also change our lived experiences through our basic human capacities.

This dynamic between the broader contextual influences as represented by patterns 1–4, the proximate influences of climate-related events, and the experiential action-learning cycle of human life enable us to produce a working conceptualisation of the lived experience of climate change. This is represented in Fig. 1.2. Note in this figure that the solid arrows indicate the main direction of influence, while the dotted reverse arrows indicate a more iterative process whereby our experiential learning may in time influence both our broader contextual influences and our capacities to deal with the proximate influences. Because our broader context also influences where we are likely to live in relation to the severity of proximate impacts, lines are also drawn between these two categories. An underlying message, therefore, is that everything is connected when we consider the lived experience of climate change.

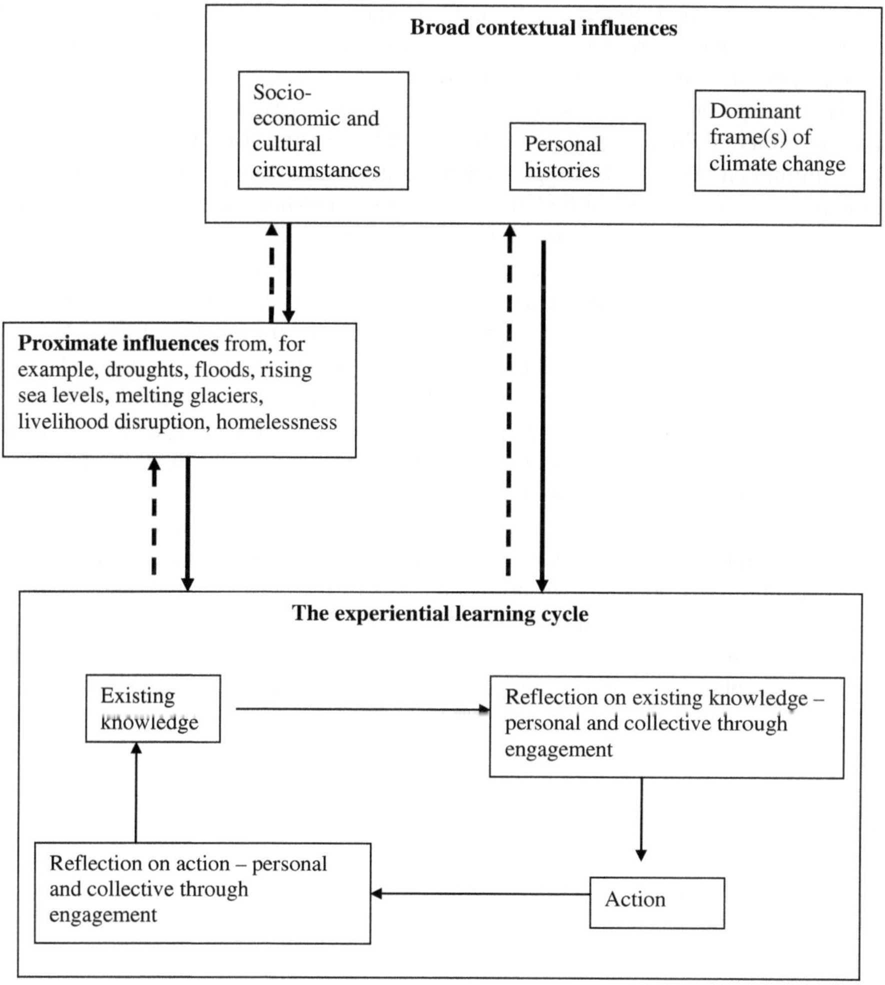

Fig. 1.2 Conceptualising the lived experience of climate change

1.5 The Building Blocks of This Book

The conceptualisation that is shown in Fig. 1.2, with all its tensions, influences and inter-connectedness, forms the first of three building blocks of our book. It runs through all chapters, but especially Chaps. 2, 4–6. Chapter 2 grounds the concept with respect to people who live in economically poor 'developing' countries and provides a deeper exploration in terms of structure and agency, concepts which approximate respectively to the broad contextual influences and the experiential learning cycle of Fig. 1.2. Chapter 4 then explores the concept in relation to how climate change is represented by disciplinary forms of knowledge from academia, and how lived experiences may enable a broader and deeper interdisciplinary understanding. Chapter 5 compares lived experience as an explanatory concept with related concepts of local and indigenous knowledge, and Chap. 6 uses it to consider more deeply the discourses of intervention—namely, mitigation and adaptation—that emanate from the IPCC.

The second building block is to examine lived experience in relation to scientific accounts of climate change. Without doubt, the latter form the dominant mode of communicating the challenge and have generated many invaluable insights. Scientists, moreover, identified the issues and communicated them in the late 1980s in such a way that the world had to take notice.

Lived experience accounts and scientific accounts, however, are likely to be as different as chalk and cheese, even if they relate to the same physical event. It is a fundamental difference that is rooted in deeply contrasting epistemologies. Yet, are they necessarily oppositional? Might they offer complementary insights? Do they separately generate knowledge that is limited by the epistemological boundaries of each? What might lived experiences offer that science cannot and vice versa? Might they combine to give a broader and deeper picture? We first explore these issues in Chap. 3 and follow up in subsequent chapters.

The third building block concerns the diversity of lived experience, and public engagement, action and policy with respect to climate change. The building block asserts itself primarily in the second half of the book, where Chaps. 7 and 8 addresses public engagement or lack of engagement with the challenge, but also the potential to develop a deeper consciousness than currently exists. Chapters 9–11 then use the platform of public engagement to discuss lived experience, knowledge, public action and policy on climate change.

References

Barton, L. (2014). A hard rain: The UK floods, six months on. *The Guardian,* 8 August. Available online at: http://www.theguardian.com/uk-news/2014/aug/08/uk-floods-six-months-on-photographs-gideon-mendel. Accessed 20 February 2015.

Blackburn, R. (1976). Okiek history. In B. A. Ogot (Ed.), *Kenya before 1900*. Nairobi: East African Publishing House.

Boko, M., Niang, I., Nyong, A., Vogel, C., Githeko, A., Medany, M., et al. (2007). Africa, climate change 2007; Impacts, adaptation and vulnerability. In M. L. Parry, O. F. Canziani, J. P. Palutikof, P. J. van der Linden, & C. E. Hanson (Eds.), *Contribution of working group II to the fourth assessment report of the intergovernmental panel on climate change*. Cambridge: Cambridge University Press.

Collins, K. (2009). Kenya says makes climate effort before Copenhagen. Reuters. Available at: http://www.reuters.com/article/2009/11/11/us-kenya-environment-idUSTRE5AA2VS20091111.

Dolfman, M. L., Wasser, S. F., & Bergman, B. (2007). The effects of hurricane Katrina on the New Orleans economy. *Monthly Labour Review, 130*(6), 3–18.

FAO. (2011). *State of the worlds forests 2011*. Rome: Food and Agriculture Organisation.

Fast, S. (2013). Social acceptance of renewable energy: Trends, concepts, and geographies. *Geography Compass, 7*(12), 853–866. doi:10.1111/gec3.12086.

Government of Kenya. (1933). *Kenya land commission evidence*. Nairobi: Government Printers.

Gregersen, H., El Lakany, H., Karsentry, A., & White, A. (2010). *Does the opportunity cost approach indicate the real cost of REDD+?*. Washington, D.C: Rights and Resources Initiative.

IPCC. (2007). Climate change 2007: The physical science basis. In S. Solomon, D. Q. M. Manning, Z. Z. Chen, M. Marquis, K. B. Averty, M. Tignorand & H. L. Miller (Eds.), *Contribution of working group I to the fourth assessment report of the intergovernmental panel on climate change. Core writing team*. Cambridge, United Kingdom, New York, NY, USA: Cambridge University Press.

Iversen, R. R., & Armstrong, A. L. (2009). Hurricane Katrina and New Orleans: What might a sociological embeddedness perspective offer disaster research and planning? *Analyses of Social Issues and Public Policy, 8*(1), 183–209.

Kabananukye, K. (2011). Ethnic–racial minorities: Movement towards political inclusion. In H. Majamba (Ed.), *Towards a rights sensitive East Africa community. The case of ethnic and racial minorities. The case of the Batwa in Uganda*. Kampala: Fountain Publishers.

Kabananukye, K., & Wily, I. (1996). *Report on a study of the Abayanda pygmies of South West Uganda for Mgahinga and Bwindi impenetrable forest conservation trust, 116*. Kampala: Mgahinga and Bwindi Impenetrable Forest Conservation Trust.

Kameri-Mbote, P., & Nyukuri, E. (2013). Climate change, law and indigenous peoples in Kenya: Ogiek and Maasai narratives. In A. Randall & E. A. Kronk (Eds.), *Climate change and indigenous peoples. The search for legal remedies*. Cheltenham: Edward and Elgar.

Kenrick, J., & Lewis, J. (2001). Discrimination against the forest people ('Pygmies') of Central Africa. In S. Chakma & M. Jensen (Eds.), *Racism against indigenous people*. Copenhagen: IWGIA.

Kolb, D. (1984). *Experiential learning. Experience as the source of learning and development*. New Jersey: Prentice Hall.

Liu, A., & Plyer, A. (2008). *The New Orleans index*. Washington, D.C: Brookings Institution; New Orleans: Greater New Orleans Community Data Center.

Muro, M., Liu, A., Sohmer, R., Warren, D., & Park, D. (2005). *New Orleans after the storm: Lessons from the past, a plan for the future*. Washington, D.C: Brookings Institution.

Mwayafu, D., & Peskett, L. (2009). *REDD-plus in Uganda: Are existing approaches' to benefit sharing a challenge?* http://www.forestcarbonasia.org/other-publications/redd-plus-in-uganda-are-existing-approaches-to-benefit-sharing-up-to-the-challenge/. Accessed 13 November 2014.

Nyukuri, E. (2013). *Climate change, policy and vulnerability of tropical forest communities in East Africa*. Ph.D. Thesis, Milton Keynes, The Open University.

Ochieng, R. (2009). *A review of degradation status of the Mau forest and possible remedial measures*. Munich: GRIN Verlag.

Ohenjo, N., Willis, R., Jackson, D., Nettleton, C., Good, K., & Mugarura, B. (2006). Health of indigenous people in Africa. *The Lancet, 367*(9526), 1937–1946.

Plaisance, S. (2006). A jump on the school year. *The Philadelphia Inquirer*, August 8.

Quarantelli, E. L. (2005). A social science research agenda for the disasters of the 21st century: Theoretical, methodological and empirical issues and their professional implementation. In R. W. Perry & E. S. Quarantelli (Eds.), *What is a disaster? New answers to old questions* (pp. 325–396). Philadelphia: Xlibris.

Stern, N. (2010). *Blueprint for a safer planet: How we can save the world and create prosperity*. London: Vintage Books.

Sung, J. (2004). Land rights: A central issue in conflict resolution and management among the Ogiek of Kenya. In V. Tauli-Corpuz & J. Carino (Eds.), *Reclaiming balance*. Tebtebba Foundation. Available online at: www.tebtebba.org.

Toulmin, C. (2009). *Climate change in Africa*. London: Zed Books.

Tove Action Group. (2014). *Reasons to oppose the wind farm.* http://www.toveactiongroup.org/reasons-to-oppose. Accessed 20 February 2015.

Tumushabe, G., & Musiime, E. (2006). Living on the margins of life. The plight of the Batwa communities of South Western Uganda. ACODE Policy series, No. 17.

Turyatunga, E. (2010). Social exclusion, marginalisation and the threat of extinction of ethnic minorities. A case of the Batwa community in Uganda. *OIDA. International Journal of Sustainable Development, 2*(3), 57–64.

UNDP. (2014). *Table 1: Human development index and its components*. United Nations Development Programme. http://hdr.undp.org/en/content/table-1-human-development-index-and-its-components. Accessed 20 February 2015.

Chapter 2
Exploring the Lived Experience of Climate Change

Abstract This chapter extends the conceptual and theoretical arguments presented in Chap. 1. It introduces lived experience as a rich and complex narrative, set within societal structures which constrain and/or enable individuals to make sense of climate change. In questioning 'whose knowledge counts' from this vantage point of view, we suggest that lived experiences of climate change provide insights and knowledge that go beyond scientific or academically presented knowledge. This experiential knowledge evolves through historical processes and is shaped through a variety of social contexts, both general and specific, between groups (Northern, Southern, rich, poor) and individuals (often defined by race, gender). It is also shaped by our personal and collective positioning in society and the scale of events that affect us. The valuable insights and the diversity can only add to existing knowledge of climate change and influence policy and practice in an inclusive way. In this, the chapter focuses on lived experiences in developing countries, the large majority of whose populations are poor and where climate change issues are linked to and subsumed within embedded poverty.

2.1 Introduction: Lived Experiences as Rich, Complex Narratives

Chapter 1 laid the foundations for thinking about lived experiences as an analytical category and introduced case studies of diverse experiences of climate change arising from both rich and poor countries. The chapter also offered a way to conceptualise lived experience of climate change by reviewing broad contextual influences, proximate influences and experiential learning cycles (Fig. 1.2).

This conceptualisation is extended in Chap. 2. In questioning 'whose knowledge counts', we suggest that lived experiences of climate change give us insights and knowledge that go beyond scientific or academically presented knowledge. This experiential knowledge is a social process of making throughout our lives. It evolves through historical processes and is shaped through a variety of social

D. Abbott and G. Wilson, *The Lived Experience of Climate Change*,
DOI 10.1007/978-3-319-17945-2_2

contexts, both general and specific, between groups (Northern, Southern, rich, poor) and individuals (often defined by race, gender). It reflects a continual tension between power relations associated with these contexts and our individual agencies to act. It is also shaped by our personal and collective positioning in society, norms and values that are associated with our culture, and the scale of interacting events that affect us. The valuable insights and the diversity can only add to existing knowledge of climate change and influence policy and practice in an inclusive way. In this, the chapter focuses on lived experiences in developing countries, the large majority of whose populations are poor and where climate change issues are linked to and subsumed within embedded poverty.

Lived experiential knowledge is by its nature subjective and arises from our long-term and short-term experiences of our encounters and dealings with the world around us. It is the knowledge we often associate with wisdom and wise sayings arising from collective or individually observed personal experiences. It is knowledge which we use to recall events in our lives so we do not repeat mistakes or help to improve our lot in life—not that we always follow the path that might appear to others who observe us to be the logical derivation from our lessons!

Lived experience is therefore personal, subjective knowledge with its own logic. It is not something we can necessarily articulate to others (or even wish to do so if the matter is very emotive). Such knowledge is often internalised, drawn from the broad spectrum of 'experiences' which of course fall back on several differing bundles of 'experiences' and encounters experienced that perhaps only make sense to the individual (but may or may not to others).

Lived experience is about our impressions, imaginations and what we hold closely and guard, but do not/can not always articulate. Our lived experiences allow us to mentally and quickly compare situations and scenarios. We carry these with us to put up self-defences, minimise perceived harm to ourselves and learn to mitigate and adapt to present circumstances. We draw on lived experiences to enhance a tacit knowledge that helps us to function every day and socialise within our particular settings.

Lived experience is then the reality of our life world, the likes of which health disciplines such as psychology and psychiatry rely on to treat patients. Other disciplinary strands such as women's studies and feminism actively encourage incorporation of lived experiences to bring out histories which have been hidden, or to discuss societal taboo subjects of sexuality. Equally, race is an important determinant shaping an individual or collective's real lived experiences which disciplinary strands such as whiteness studies draw out in order to challenge historical assumptions. Again, how much race has contributed to personal experiences cannot always be neatly articulated or codified.

It is clear then that lived experience is not a 'science' and goes against the grain of natural science which sometimes appears distant and baffling to the lay person. Equally natural scientists may query the rigour of lived experiences using scientific criteria because they cannot be neatly articulated. Nevertheless, discovering and representing individual and shared experiences may open up realities of a life world that are beyond statistics and numbers. For instance, development studies, which purport to be concerned with poverty, justice and inequality, from local to international scales, make significant use of 'lived experience' to understand the realities of poverty.[1] Whilst financial and other powerful global institutions use several bland but functional economic indicators such as Gross National Product (GNP), Gross Domestic Product (GDP) and inflation measures to indicate levels of national poverty, they do not say much about its reality to the individuals and people of these countries.

For example, the eight United Nations Millennium Development Goals (MDGs) were agreed at an international Millennium Summit of world leaders in 2000, to be reached by 2015.[2] Each Goal is accompanied by specific targets. Thus, Goal 1, Target 1.A reads: 'Eradicate extreme poverty and hunger. Halve, between 1990 and 2015, the proportion of people whose income is less than $1.25 a day'. Arguably, this target aims to take into account individuals and their families. The $1.25 criteria by any reasoning, however, remains a one-dimensional reductionist measure providing the world with no account of how the poor juggle, multi-task, borrow, beg, steal, dodge moneylenders and violence to manage survival on that amount of money. Neither does it tell us how this differs for men, women and children and the powerlessness that they may experience in poverty.

In contrast to the measure that defines poverty as living on less than $1.25 per day, Development Studies research on 'lived experience' is used to capture the subjective, contextual, multi-dimensional nature of poverty as it affects real lives. To make a systemic analysis and sense of what is told (often shouted!) by these voices of the poor, special participatory methodologies have emerged, often comparing and emanating from other humanistic science disciplines. The complexity and richness of 'lived experience' that are derived from poor people's narratives in these participatory settings provide a story of poverty which allows researchers to consider the varying degrees of power which shape an individual's position in society, between groups differentiated by class, caste, gender, ethnicity, race and religion.

Whilst it is easy to articulate scientifically the numbers in poverty associated with 'less than $1.25 a day' and irrespective of whether MDG Goal 1 targets have been achieved or not, the narratives of lived experience are problematic to codify even if qualitative methodology allows this to an extent. As illustrated in Chap. 1, it is, however, possible to decipher some patterns that act to impoverish individuals, communities and even countries. It is important to discover these patterns (see Chap. 4 for more on pattern making) and what creates them if we are to break cycles of impoverishment and look at the cause rather than a descriptive measure of poverty. Thus, lived experiences add to our knowledge of poverty and survival and complement statistical macro-pictures, which is why it is important to focus on them.

[1] Poverty and climate change are inexorably linked. This is further detailed in Sect. 2.4.

[2] There is much debate as we go to press regarding the extent to which individual MDGs have been, or are being, met. Important as this debate is, it is tangential to the focus of the book and we do not cover it here or elsewhere.

Thus although lived experience relies on individual and collective narratives, as we explore below, it is possible to articulate commonalities and patterns that can inform us. Lived experiences also have the capacity to represent some commonalities in relation to climate change, but because of the diversity of these experiences, they do not necessarily represent all. Yet these experiences can be shared and certainly give us a picture of context, scale and events that are often difficult to imagine through a science-dominated issue such as climate change. To consider how a systematic and systemic analysis of lived experiences may take shape (and therefore be articulated), we can think of them as 'social processes of making'—influenced by personal and collective historical narratives, in turn influencing and making societies.

2.2 Lived Experience as a Social Process of Making

To articulate lived experience as a social process of making, there needs to be at least a two-fold consideration: (i) how the lived experience evolves over time and (ii) how it is shaped by the interplay between 'structure' and 'agency' (Box 2.1), both of which require a systematic study of society and societal relationships.

Box 2.1 Concepts of structure and agency

Structure The pattern or framework of relationships between social institutions such as markets, families, class, and political factions. It includes rules of behaviour associated, for example, with moral norms and hierarchies.

Agency: The actions of individuals or groups, and their capacity to influence events.

In practice, social analysts combine both concepts, although some tend to favour one over the other. (*Source* Allen and Thomas 2000: 189.)

For climate change, as indicated by Fig. 1.2 in Chap. 1, the interplay takes place between the 'external' influences (biophysical and socio-economic contexts) and our conscious actions to do something about the situation. Another way of examining this is to consider the extent to which we consciously create our lived experiences through 'free will' and the extent to which our experiences are determined by contextual factors. Lived experience of climate change also, of course, works in a similar way, evolving over time, incorporating individual and collective knowledge set within diverse contexts of external influences and internalised understanding.

As we explore further in Chap. 4, the study of society and societal relationships spans a wide range of social science and humanities subject disciplines that are based on patterns of human interaction. These include, for instance, various strands of sociology, economic, cultural, political and religious life, both at a local and global scale, and analyses of the complexities of human interaction from micro- to macro-levels. Such a study in relation to climate change, for instance, would allow at least a three-way analysis: (i) of individual behaviours and attitudes, (ii) set within specific local societal contexts, and (iii) within a global context reflecting global structures, issues and actions.

An analysis of individual behaviour requires a critical understanding of social structures within societies, whether in the immediate locality of the individual or within a worldwide context. Much of its theoretical and conceptual analysis is thus around deciphering the everyday relationship between these contexts (structures) and individual behaviour (agency) as noted in Box 2.1.

In these interactions, our daily lives and interaction with each other are patterned, and generally revolve around various organisations, institutions and structures in our immediate lives and wider society in a predictable manner (Box 2.2). These predictions and patterns are defined by the expectations of the roles and status that each of us have within society and are important to its functioning—otherwise interaction in a random manner is often counterproductive and confusing! As humans, therefore, we require relative stability to function every day and the framework for this is located within structures embedded in social institutions and organisations which are an essential part of any society.

Box 2.2 Institutions

Institutions are closely related to organisations and structures. The later Chap. 10, Sect. 10.1 points out those institutions can also be organisations but not all organisations are necessarily institutions. This is because institutions define formal rules which can be enforced (e.g. by state-backed government, police, army institutions), informal expectations (e.g. through social institutions of marriage, family, church) and dualistic which operate both formally and informally (e.g. by educational institutions such as schools and universities).

Cultural theorists, feminists, anti-racists have shown that there is a strong connection between differing contexts and entrenched institutional rules and practices. Challenges to institutional rules and practices also challenge dominant ideologies—representing shared political and philosophical values and attitudes usually of the most powerful and influential people—in turn raising much awareness and action in generating institutional change. For example, (although we are not quite there yet) segregation based on race in the American South, in South Africa and other places has brought about changes in institutional rules and practices. Gender inequality of opportunity has been challenged in many parts of the world within marriages, domestic and work contexts, in turn influencing patriarchal ideologies.

Institutional rules, whether formal or informal, represent power within social structures and are constantly being battled over. We can, in this battle, go forward some steps (e.g. with women's rights to education) but equally there are those who will fight to revert back to rules that suit them (for example, denial of female education by fundamentalist groups). Institutions are therefore a battleground of power politics played out in full public view.

These structures include families (represented through the institution of marriage), religion (represented through churches, mosques, etc.), culture (through everyday norms, practices and values), education (through schools, universities) and organisations (through employment, official bureaucracies). Such structures are crucial within our everyday lives which they shape and hold together. In fact, the importance of these structures may be compared to the rafters and steel girders that hold up buildings, supporting the overall frame upon which the foundations and future construction of the structure are based. At the same time, the rafters and girders provide an overall, overarching influence on the construction and overall future functioning of the building!

To articulate lived experience as a social process, it is, of course, important to be interested in what is upfront (i.e. the building), but more fascinating is what underpins societal behaviour (i.e. the rafters and the girders). It is therefore critical to identify those often taken-for-granted and sometimes invisible structures that are most influential in shaping human interaction in relation to various social issues (of which climate change is one). As societies have many layers through which both individual and societal behaviour is defined, understanding can only come through unpacking these systematically in order to decipher what lies behind individual behaviour.

The argument is that whilst the individual is important, the society within which the individual lives and functions is even more so! This relation between the individual and society, however, is not always immediately apparent as individual behaviour is in fact often internalised or aspects of it are given, or hidden from a public view. Lived experience is a social process which evolves over time arising from the specific historical, economic and social context of which we are creatures. The general consensus which includes scientists, non-scientists and the wider civil society is that climate change is a hugely complex problem. This requires multifaceted problem solving and systems thinking as summarised in Box 2.3, where climate change is viewed as a problem which is part of an overall system, rather than specific bits which need to be resolved (for example, carbon emissions), outcomes (for example, global warming) or events (for example, flooding from extreme weather).

Box 2.3 Systems thinking

This is the process of understanding how things influence one another within a whole. In nature, systems thinking examples include ecosystems in which various elements such as air, water, movement, plants and animals work together to survive or perish. In organisations, systems consist of people, structures, and processes that work together to make an organisation healthy or unhealthy.

Systems thinking has been defined as an approach to problem solving, by viewing 'problems' as parts of an overall system, rather than reacting to specific parts, outcomes or events and potentially contributing to further development of unintended consequences.

Source http://en.wikipedia.org/wiki/Systems_thinking (accessed 27 August 2014).

Currently, there are several critical issues we face as human beings (including a future of food and water shortages) and further unintended consequences of climate change. To complement existing knowledge, lived experience adds a dimension that examines internalised experiential and how societies draw on it to mitigate and act. One way of decoding this knowledge through the overall processes of societal 'systems' is by unpacking the complex relationship between individuals and structure through power relations that shape peoples' lived experiences making it possible to analyse these and articulate how people shape their understandings of climate change.

2.3 The Complex Interaction of Structure (Power Relations) and Agency that Makes Lived Experience

Arguably, it is the complex interaction between societal structures and their differing levels of power relations and individual agency that shape and make our lived experiences. The debate over which—structure or agency—is most influential goes back a long way including the Bible and the ancient Greek philosophers. To step to a closer period, informed by Darwinian evolutionary theories and Durkheim's sociological analysis (Chap. 4), sociologists, psychologists and philosophers have been fascinated by the discourse on whether individual behaviour emanates from heredity 'nature' and the biological characteristics people are born with, or the acquired ones from their 'nurtured' social environmental contexts. Although focus on the nature/nurture explanations of individual behaviour continue to differ, generally (and in common sense), they are not juxtaposed as opposites, but 'a bit of both'.

Equally structure/agency arguments also go over these well-worn routes in questioning what shapes our life experiences, opportunities and behavioural responses. For instance, the poverty example referred to in Sect. 2.1 could be linked to consequences of structural poverty in two ways: (i) through the overarching country poverty and low GDP/GNP per capita in a globalised world, and (ii) through individual stories caught up within the power and unequal relationships of a local societal context which constrain and determine choices. Of course, they could also be linked to agency choices of 'free will'—that is, making bad agency choices in life. Lived experiences are a result of the complex interaction between these national and individual societal hierarchies, the individual's immediate world and that which goes beyond that—shaping the choices they make within these circumstances.

Structure/agency debates attempt to unpack the interaction between individuals (agents) and society (structures) not as separate, compartmentalised categories, but as interconnected and interdependent. The argument, however, is about the emphasis and the focal role of one over the other, very simply put, (i) that individuals are rational and can exercise agency to overcome societal and institutional barriers and seek opportunities that will enhance their life chances, and (ii) that starting from birth, societal institutions constrain life chances so that individual agency is limited. Societal power relations and who has power and why are issues pivotal to these arguments and are reflected in policy making on equality.

Over time, sociologists have made the structure/agency debates into a fine art. Giddens (1993), for example, refers to a 'duality of structure' in arguing that individuals are constrained by society, but at the same time they are active in making it. The two are inexorably interlinked in his suggestion that 'social structures are both constituted by human agency, and yet at the same time are the very medium of this constitution' (Ibid: 169). In adapting a phrase from Karl Marx (Chap. 4) over 100 years earlier he does, however, focus on the importance of structure in that, 'The realm of human agency is bounded. Individuals produce society, but they do so as historically located actors, and not under conditions of their own choosing' (Ibid: 168).

Others, whilst recognising the importance of structure, also draw on individual agency as well. For example two other sociologists, Berger and Luckmann (1967) suggest that reality, as understood by the individual, is in fact based on the societal structures within which each person interacts. This argument may appear in the first instance as structurally bound in that individuals are seen as limited or constrained by societal structures and social forces which ultimately influence and therefore define the individual. However, in developing their argument, both Berger and Luckmann are also careful to acknowledge the importance of individuality, individual uniqueness and the ability of some to exercise a degree of independent agency.

To understand the interaction between structure and agency and how these reflect power relations, it is useful to look at general social patterns as well as particular individual actions and behaviours within specific social set-ups. Thus an individual might have particular characteristics which define them (e.g. entrepreneurship, determination) which enable them to transcend social boundaries, but essentially it is the general social discriminators of age, gender, and social class which have a bearing on their behaviour and life chances. Moreover, this becomes an especially strong bearing in poor communities of developing countries which provide our examples for this chapter. Lived experience is a social process which evolves and emanates from this interaction and is embedded in the individual's knowledge of how to manage any social constraints.

Take for example, small women farmers in The Gambia, a small ex-British colony in West Africa, who are often very poor, and responsible for feeding their families. Many are also responsible for domestic, subsistence cropping and attempt to grow the staple, rice, in swamp water. These women farmers usually work very hard carrying out the various processes of 'stoop' labour required in rice farming, namely sowing, transferring, weeding and harvesting the crop.

Disentangling what is happening at the societal level, however, reveals a number of structural limitations and ideological controls that combine to force the women to work so hard. For instance, the women rarely own their own land and are pushed onto the margins where their access to land is limited to the swamps which do not have the most fertile and advantageous characteristics for rice growing. The end crop is therefore never sufficient and women are forced to enter into seasonal, waged labour, relying on male farmers who may own land. They may also diversify into other activities such as vegetable growing. Here too they are limited to doing so within the boundaries of the household compound land.

Essentially, therefore, no matter how hard the woman works or how determined she is, in a heavily patriarchal society where decision-making is in the hands of men, the structural barriers which deny her land ownership curtail her chances of making an adequate livelihood. She also remains threatened by the polygamous practices of marriage in The Gambia where even if co-wives do cooperate in some matters (e.g. looking after each other's children on occasion), she is ultimately responsible for raising and educating her own children rather than the father or the family group.

This example is, however, only that of a localised context. At a wider level, global markets, trading and inequality between sovereign nations play an important role in determining monetary values of local currencies, spending and earning power of individuals and overall national poverty and food crisis. The ideological context within which global societies operate is also decisive in analysing the relationship between structure and agency. For example, in free-markets led by 'western' societies where individuality and individual entrepreneurship are particularly valued, the notion that our lives are shaped by predictable, patterned behaviour determined by social structures is not something that is readily acceptable to many. It is of little wonder than that many would 'blame' the women farmers in The Gambian example for not being able to farm rice (the staple food) efficiently, thus creating dependency on external sources, or peanut cash-crop farmers for not being efficient or competitive enough to produce or process a crop that competes with giant, highly mechanised American peanut butter production. Is it the farmer or the ultimate power play of global inequalities that are to blame? This, of course, depends on the nature of the structures and their constraining capacity and whether or not it is almost impossible to break free from them? Thus, the structures that constrain Gambian women rice farmers are huge, those that constrain American peanut butter producers less so—even those who are relatively poor in relation to American society.

'Power' and 'powerless' embedded in social structures shape individual actions. Power is central to the understanding of everyday social relations and enables us to ask who the powerful are, what enables them to exert power, and how they influence individual behaviour. Such power is not always apparent and the individual may or may not be aware of who holds it. The small woman farmer in the Gambia, often illiterate, is unlikely to be aware of the power of rice corporates or the buying of land by the Chinese in a renewed, post-colonial scramble-for-Africa. They may also internalise the power relations closer to home in not questioning polygamous practice and male domination in determining all life chances.

Critical determinants such as money, the class or caste you are born into, the race and gender that you belong to, more often than not influence how powerful (or powerless) you are within a given society. The more powerful you are, the more you are likely to influence the dominant structures of your society (for example, through religious, education, and family institutions). As political, ethical or moral beings, it is important to question how the shift from 'power' to 'empowerment', and conversely the related issues of 'powerlessness' and 'disempowerment' occurs, if it ever does. This requires an analysis of individual actions and how these emanate from, and result from interaction with, other individuals in the wider society. In turn, this necessitates a deeper, below the surface, exploration of structural equality and inequality.

There is little doubt that some individuals can successfully challenge the power embedded in social structures—fighting for example strong apartheid policies (Nelson Mendala), overt historical American racism (Barrack Obama); fierce castism (Dr. Bhimrao Ramji Ambedkar the 'untouchable' author of the Indian constitution) to name a few. Nevertheless, behind each successful individual from the minorities, there is a legacy of discrimination suffered by millions, clearly reflecting that most individual life experiences remain bounded by structural power relations of racial, caste and class hierarchies.

Climate change stories also require a structure/agency power relation analysis. For instance the use of fuel wood by the poor is often associated with significant levels of local deforestation leading to climatic consequences in developing countries (Agarwal 1986, 2010). People in many developing countries use fuel wood for two main purposes (i) domestic, such as for cooking, and (ii) for marketing and trading purposes. Depending on the number of people who cut down forests to obtain the wood, deforestation often occurs fairly rapidly and soon starts to become visible. Yet people go further and further into the forest to obtain the wood, causing increasing pressure on the forest.

Arguably, in this example, the individual has choices and can exercise agency on whether to cut the wood or not. They may not want to do so as often poor, landless people (particularly women whose responsibility it is to gather wood) rely on common resources such as forests and are very aware of their value and the cost of their deterioration and loss. Yet, especially those in developing countries carry on cutting wood because they have limited other resources.

To understand how individual lived experiences evolve over time and the interaction of structure and agency that make them so, requires a 'sociological imagination', a term associated with C. Wright Mills, an American sociologist who in 1959 suggested that there is a relation between individual biographies and historical change embodied in global and local societal events. A 'sociological imagination' is required to transcend the epistemological boundaries of everyday experience in order to establish the relationships to which he refers, and therefore exhibit a broader, challenging agency.

A fuller argument on this relationship is represented in Box 2.4. Mills demonstrates that even if we do not acknowledge it, or understand it, our individual biographies are not simply about ourselves or how we shape them, but an interplay of our individual selves and historical change that is embedded and manifest within societal structures as described at the start of this section, and well illustrated in the second paragraph of the extract. In Chap. 8, we develop further Mills' concept of 'sociological imagination'. With respect to climate change, we re-conceptualise it as a transboundary, social imagination that is generated through engagement of diverse lived experiences and between lived experiences and the sciences.

One way to assess this interplay is to consider how historical processes and change affect individual responses. Thus, for instance, India is undergoing transition from what was essentially a closed society based on internal markets towards trade liberalisation and wider engagement within global markets. In turn this has generated institutional, ideological and structural change which can enable businesses and entrepreneurs to thrive. Opening up to external global markets has also created a significant group of rich and super-rich Indians, and amongst these, a large number of rising middle classes whose behaviour has changed in that they are more demanding of 'western' style consumption such as large cars, fridge/freezers, gadgets and are acquiring a 'throw-away' attitude that is challenging to the environment. In fact, today in India this has raised many controversies regarding the effect of such large-scale consumption with respect to climate change issues.

Box 2.4 On interaction between individual biographies and historical change

Nowadays people often feel that their private lives are a series of traps. They sense that within their everyday worlds, they cannot overcome their troubles, and in this feeling, they are often quite correct. What ordinary people are directly aware of and what they try to do are bounded by the private orbits in which they live; their visions and their powers are limited to the close-up scenes of job, family, neighbourhood; in other milieu, they move vicariously and remain spectators. And the more aware they become, however vaguely, of ambitions and of threats which transcend their immediate locales, the more trapped they seem to feel.

Underlying this sense of being trapped are seemingly impersonal changes in the very structure of continent-wide societies. The facts of contemporary history are also facts about the success and the failure of individual men and women. When a society is industrialised, a peasant becomes a worker; a feudal lord is liquidated or becomes a businessman. When classes rise or fall, a person is employed or unemployed; when the rate of investment goes up or down, a person takes new heart or goes broke. When wars happen, an insurance salesperson becomes a rocket launcher; a store clerk, a radar operator; a wife or husband lives alone; a child grows up without a parent. Neither the life of an individual nor the history of a society can be understood without understanding both.

Yet people do not usually define the troubles they endure in terms of historical change and institutional contradiction. The well-being they enjoy, they do not usually impute to the big ups and downs of the societies in which they live. Seldom aware of the intricate connection between the patterns of their own lives and the course of world history, ordinary people do not usually know what this connection means for the kinds of people they are becoming and for the kinds of history-making in which they might take part. They do not possess the quality of mind essential to grasp the interplay of individuals and society, of biography and history, of self and world. They cannot cope with their personal troubles in such ways as to control the structural transformations that usually lie behind them. (*Source* Mills Chap. 1, The Promise page 5, cited from the 1959, 1st ed. and Mills 2000, 40th ed.).

Lived experience therefore evolves out of a complex interaction of power relations bound within structures of society where each individual may exercise varying levels of agency depending on their personal power positions. To deconstruct these relationships, a sociological imagination allows us to 'think oneself away' from the familiar routines of day-to-day life. It is only when we explore beyond the familiar patterns of everyday life that we begin to discover what underlying forces shape our behaviour as individuals. The process of uncovering the deviant thus ignites imagination further, and becomes even more alive as it allows us to consider what is happening not only in the mainstream, but on the margins of society and makes for a more inclusive account.

2.4 Lived Experience in Relation to Interacting Phenomena

As suggested in the section above, lived experiences are systemic, that is, they are made up of our encounters within the various interacting events and happenings that shape our personal histories and understanding of life. Climate change, for instance, brings differing experiences and reactions depending on whether you are located in affluent or developing countries, rich or poor, even though environmental degradation may affect us all.

Looking generally at the macro picture between the impact of climate change on rich and poor countries across the world, it has become apparent that poor countries are more vulnerable. Mendelsohn et al. (2006), for instance, argue that even if technology and resources should be directed equitably across countries to mitigate the consequences of their unequal distribution, other geographical factors come into play. For example, some poor countries, with low latitudes whose baseline temperature is already very hot, are finding that further warming results in pushing them even further away from the optimum, making them directly prone to climate change. Frankhauser and McDermott (2014) suggest an important reason why poor countries in comparison to rich countries are currently affected so adversely by extreme weather events and will continue to do so is because of an 'adaptation deficit'. In general terms, there is a positive association between income and ability to adapt to climate change, and conversely a limited ability of poorer countries to adapt.

At an individual and household micro-level in developing countries, it is the vulnerable, especially women in poverty who are more likely to be threatened by climate change. For example, Irene Dankleman (2011: 1) quotes one woman:

I have seven children. [...] Our house is collapsed because of the floods, and our crops—maize and sorghum—are destroyed. Hunger stares us in the eyes. Mrs. Atibzel Abaande, 45 year, Bawku West District, Ghana.

She then comments:

> For billions of women and girls in this world there are less [sic] possibilities, opportunities and certainties than for many men. More and more we see that changes in biodiversity,

environment and climate have an increasing effect on such gender inequalities. On the other hand: if women and girls get opportunities, they have proved to be important actors on the way to a more sustainable and just society.

Weather extremes and chaos that are associated with climate change are not gender-neutral but do affect women in poverty the most. They are particularly vulnerable all over the world because of several underlying commonalities which manifest themselves in powerlessness for individual women in their everyday lives. These include:

(i) Poor women are less likely to own land or property yet are more dependent than men on natural resources for their livelihoods. Thus as 'free' (communal access) natural resources diminish, their chances of making a living are also threatened;
(ii) Their power in society is limited by social, economic and political structures that constrain their capacity to cope, particularly if they are single, widowed, abandoned heads of households looking after their own or other children;
(iii) They are often directly responsible for finding basic fuel, food and water supplies for their family in spite of limited mobility, particularly in the rural areas. Problems intensify when changes occur to essential resources such as water salinity, which accompany climate variability especially in coastal areas. When nearby potable water is affected, women have to travel even further distances often on rough terrain in search of alternative supplies (WEDO 2008).

Women continue to suffer large-scale social inequalities in terms of both material and non-material matters (see Branisa et al. 2014 for a fuller index of the institutional basis for gender inequality). The former include inequalities in land and property ownership as in (i) above, and inheritance rights. The latter include issues of decision-making within households, reproductive health, education and general human and political rights.

Thus, individual actions and reactions to climate change are shaped according to women's age, gender, class and other discriminatory factors. The local structure is, of course, embedded within the global context. For instance, we are increasingly aware of the global interconnectedness of biophysical climatic impacts, such as evapotranspiration and desertification. Changes to climate and climate-dependent biophysical environments are nowadays hard facts of everyday reality, but they are increasingly a pressing issue for developing countries throughout the world. In fact, it seems almost inevitable that the more economically poor a country, the more it is likely to be affected adversely as it has a lower capacity to cope with this change (IPCC 2007).

It is important, however, to remember that neither the poor generally nor women in poverty specifically are passive victims. A very strong global fight-back is evident in the way social movements around green issues have evolved. These include, for example, the Environmental Justice movement which began in 1982 in the economically developed world. Hundreds of activists and local residents began to protest against the expansion of a chemical landfill in poor communities of Western Country, North Carolina in the United States. The movement has since opposed and stopped many plans to turn communities into 'waste incinerator dumps' and 'toxic doughnuts' where communities are surrounded by a full circle of toxins (Pellow and Brulle 2009: 435). Another historical example is from a developing country where the 'Chipko', tree-hugging women led a huge protest against the Indian government and private firms wanting to destroy forests in the North of India in order to make profit from logging—a movement that has now spread to other countries (Shiva 1989: 67). We discuss social movements as a countervailing force around climate change issues further in Chap. 11.

Through lessons learned from disasters such as floods, hurricanes, and the predictions made by climatologists, it is clear that social disintegration and increasing inequality accompanies climate change. As vulnerability increases, societal chaos results, exemplified by mass migration, disruption of kinship and peer support systems, displacement, loss of subsistence economies, conflict and warfare. Examples are found in instances of (i) migration arising from the recent drought and related conflict which have afflicted thousands from Somalia to Kenya, or (ii) with the millions who try to escape rural poverty, which has been intensified by environmental changes, in South America, Asia and Africa, by joining the ever-growing numbers in mega-cities (Raleigh et al. 2008). Therefore, the consequences of climate change and increasing inequality are far-reaching, not only across immediate neighbourhoods and localities, but also across wider geographical spaces and boundaries. It is important to note, however, that we (the authors) do not just attribute this bleak future scenario to climate change alone as we are always concerned with a multi-variable rather than a single variable explanation.

There are complex, interacting links between climate change and social phenomena of inequality, and its counterpart, equality. Arguably, climate change is also capable of providing new opportunities to bring nations and individuals together to create equality which will ultimately challenge existing inequalities. This is because in climate change everyone, rich and poor alike, is affected and we cannot escape our interdependence on the challenge that faces us. The sociologist, Ulrich Beck, who has brought the issues of societal behaviour and future global sustainability to the forefront, for instance, argues:

> Climate change globalises and radicalises social inequality; it exacerbates inequalities of rich and poor, core and periphery, and at the same time dissolves them in the face of a common threat to humanity. Climate change combines with the inequalities arising from globalisation, decoupling the producers and subjects of risk. Remapping inequality in the age of climate change and globalisation therefore requires taking account of the unbounding of both equality and inequality, and an awareness of the end of the opposition between society and nature, one of the founding principles of sociology. (Beck 2010: 165).

He further argues (ibid) that whilst climate change has the potential to increase gaps and lead to social disintegration, a move away from nationalist principles and solidarity has the potential to create a global transnational solidarity in the face of climate change scenarios that impacts all humans. Uniting to tackle this has the potential to close divides and create global equality.

Climate change lived experiences can therefore be understood in relation to interacting phenomena of poverty, social positioning, equality and inequality even if both the rich and poor are affected. See Chaps. 10 and 11 for further exploration of 'interdependence'.

2.5 Lived Experience in Relation to Culture and Value

The significant role that cultural norms and value systems play in shaping personalities, societies and action has long been recognised in several diverse disciplinary strands of academia, including anthropology, psychology and health. Regarding climate change, it has become increasingly critical to communicate effectively with the public, where there is an expanding associated literature on the need to understand the critical role of human values and worldviews, the latter being the ways in which a specific society views and articulates its engagement with the world around it. (For a fuller review of the literature on this topic, see Corner et al. 2014).

Cultural norms and values are inexorably linked but there is a subtle difference between the two terms. Simply put, norms (which may not always be good practices) are internalised, learnt patterns that we see as acceptable behaviour and attitudes within a given society. Values are embedded in what we see as right or wrong within our society and include morals, ethics and beliefs. Our understanding and response to every dimension of global climate change is mediated through our cultural values. For example, Adger et al. (2013) show that climate change threatens many aspects of lives and livelihoods including lived aspects of culture, identity, community cohesion and sense of place. How societies respond to these changes and adapt to climate-related risks depends on these cultural dimensions.

In considering cultural values, Wolf et al. (2013) suggest that dominant approaches to climate change have usually centred on issues of vulnerability with an underrepresented focus on local materials and symbolic values. The authors define values as 'transituational conceptions of the desirable that give meaning to behaviour and events, and influence perception and interpretation of situations and events' (Ibid: 1). To develop value frames for their research, they concentrate on the diverse and nuanced subjective, intangible meanings people attach to specific aspects of their way of life. The authors argue that 'values are crucial in shaping perception of climate impacts and adaptation to them. Distinct values, such as tradition, freedom, harmony, safety, and unity shape different interpretations and meaning of impacts, and lead to distinct views on how to adapt to these. Conflicting and competing values can act as barriers to house is collapsed because of the floodsadaptation. The findings [that is, of the study] imply that adaptation research and policy need to address values explicitly if efforts for planned adaptation are to be perceived as legitimate and effective by those affected by the changing climate (Ibid.)'. We endorse this point, while moving away from notions of linear influence and arguing for a circular relationship between lived experience and values in Chap. 11.

Disaster studies suggest that response and adaptability are located within the complex pre-existing political, social and economic circumstances that are framed by local cultural values and associated behaviours (McGilvray and Gamburd 2013; Irshad 2014). Take, for instance, the notion that disasters are gendered (Seager

2014; Zommers and Singh 2014). The Bangladesh Cyclone of 1991 exemplifies this in its death rate difference between men and women in a rigid patriarchal society both during and following the event. Chowdhury et al. (1993) show that, during the cyclone and flooding, more women than men died. The question then is why certain individuals died whilst others survived, and the underlying reasons for more female than male deaths. Were these skewed ratios a result of greater individual (biological) male capacity to challenge the cyclone through essentialist determinism and strength (that is, pure agency), or were they the result of something else embedded in the local context?

Chowdhury et al. (ibid) identify several reasons for the female: male death ratio in this particular cyclone, reasons that are primarily bound within the values and behaviours associated with societal structures and context of a Bangladeshi society, which in turn is influenced by Islamic religious values and codes of behaviour. They argue that, for a start, women were constrained by societal norms governing their dress and public behaviour which prevented them from responding to the crisis as quickly as the men. Many of the women could not swim as this is a public activity, not generally appropriate to female modesty.

They were further restricted by clothes that covered them from head to toe, with some wearing the outer veil of a 'burqa' which hampered their survival efforts. Because they were isolated and unable to move freely in public spaces, women were also reliant on men to both supply them with information as well as accompany them to a place of safety. In addition, women were not able to access public spaces as readily as the men and it took them longer to get to safety. In fact many of the women left their homes too late because they waited for a male relative to accompany them. This kind of evidence suggests that the structural constraints for women in Bangladeshi society are so strong that they are sometimes rendered powerless to exercise agency, to the point of death.

There are several other factors besides fatality that also reflect powerlessness for women in Bangladesh. Such powerlessness is exacerbated when disasters such as the cyclone occur. Flooding, for instance, puts a halt to any agricultural or livestock activity. It brings homelessness and dependency on relief. As men are also left homeless they are reported to vent their anger increasingly on women (Chowdhury et al. 1993; Ikeda 1995). Thus it is common that physical, sexual and emotional violence increases during and after a disaster whether in rich or poor countries. Japanese and Australian disaster management studies, for example show that there is often increased intimate partner violence and exploitation of single women, single mothers, and those otherwise unattached or living alone in economically developed countries (Yoko 2014; Sety et al. 2014). As Chowdhury et al. (ibid) further show, in Bangladesh, flood and cyclone shelters along with any relief centres throw women (used to sexual segregation) into sharing spaces with strangers where the men are more likely to sexually harass them. In addition, because they might find it more difficult to protect themselves, these shared spaces are particularly difficult for vulnerable women such as disabled women and nursing mothers.

Taken as a whole, it would appear that the underlying reason for greater female deaths during the 1991 Bangladeshi cyclone, as an example of a climate-related disaster, is that of the values and norms associated with the social context rather than of inherent, individual incapacity to deal with the challenge. Other research comes to similar conclusions. For example, a 2006 study of 141 natural disasters suggests that when men and women's economic and social rights are more or less equal, there is little or no difference between male and female deaths related to a disaster event. Female mortality is higher, however, where women have fewer rights than men (Neumayer and Plümper 2007). This type of conclusion can only be drawn by analysing underlying social structures (in this case, gendered behaviour of individuals) and the lens of specific cultural and value systems of the given context. Note that we also explore the notions of culture and values in Chap. 7, in relation to engagement on climate change.

2.6 The Lived Experience of Climate Change: Personal and Collective

The personal, as the above discussion suggests, is located within each individual's accumulated life experiences, and bound within their histories, societal structures, economic and political contexts. These individual accounts are important in that they enhance our understanding of specific contexts and our understanding of the reality through our place in society. This includes particularly those who are marginalised and have historically been hidden from history and denied voices such as the poor for whom changes in the climate often threaten livelihoods, and matter a lot.

Understanding how these individual accounts make up a collective is not just a matter of scaling up to larger numbers. Collective accounts of climate change usually centre around shared experiences of particular groups, coming together for mutual benefit in adapting and mitigating to threats or changed circumstances. The experiences include emotions attached to myths, beliefs, and indigenous knowledge particularly of the elders of the community/group.

For instance, the Bhils in Rajasthan, India, like many other world tribal groups worship the *Magra Baoji* (living mountain), represented by rock bones, river blood and tree and moss hair. Emotionally, they believe that the mountain also feels pain caused by mines, clearing and pollution. In a collective belief they thus oppose any action that threatens the mountain, woods and wildlife surrounding their homes (Snodgrass et al. 2013: 328).

Collective emotions often manifest in religious belief of the sanctified, play an important role in action and protest against perceived degradation and change. For instance, Hindus see the holy river (Ganges) and life-giver to many millions on whose water they depend as 'Mother Ganga'. There is a great reverence for the waters of the Ganga that goes beyond the Hindu faith and is understood by all Indians, whatever the religious denomination. This was brought home to me clearly when at a busy and overcrowded Delhi railway station a man occupied half of a much sought-after bench by placing his suitcase on it. Everyone who asked him to move it was told that it contained a jar of Ganga water (Sanskrit: Ganga jal) which he could not disrespect by putting the suitcase on the floor. To my surprise people, recognising the need for reverence to Ganga jal, accepted this immediately and quietly moved away without any further comment.

The sacred river which is so closely linked to the nation's psyche is worshipped, loved and seen as a gift from the Gods, particularly Lord Shiva whose abode is in the Himalayas. Anything that hurts the Ganga hurts Indians. Thus development projects (hydro-electric dams, deforestation) which further lead to climate shifts such as altered precipitation, warming temperatures, and melting glaciers are commonly viewed as defiling the Mother. Devotees speak of love and loss of intimacy, identity and familiarity with the waters that accompany the evident changes in the river's appearance (Drew 2013).

Equally, there are collective implications to well-being and physical health as demonstrated by Canada's Inuit populations who in the face of altering local ecosystems find it increasingly challenging to travel or hunt on the land as they have done in the past. Inuit narratives of lived experiences suggest that the changes in land, snow, ice, and weather impact their culture and self-worth, causing anxiety, sadness, depression, fear, and anger (Wilcox et al. 2013). (Note, also, however, that the Inuit are not passive as they adapt to climate change—see Chap. 5, Box 5.1).

Environmental change, however, is not always seen specifically as led by climate-related events or disasters. Most indigenous peoples also ascribe meanings to the change through other social events in the community, such as value change in the community, population growth, out-migration, urbanisation, and land degradation. A study of two Quechua-speaking farmer communities from mountainous areas near Cochabamba, Bolivia (Boillat and Berkes 2013) shows that observations of climate change are interpreted as part of a cycle, which includes a belief in the return of some characteristics of ancient or mythological times, a reaction of natural or spiritual entities that people consider living beings (as with the Magra Bhoji above). This interpretation is a particular process of lived experience which contributes towards adaptive strategies. 'Indigenous knowledge' is the term that is often used to express the manifestation of this process of lived experience, which we explore further in Chap. 5.

We end this section on personal and collective meanings of lived experience with a selection of quotations from the participants in the Boillat and Berkes (Ibid: 6) study. They show how personal observations, elder lived experiential knowledge and their links to general societal change may explain climate variations for a group of people who probably cannot access academic scientific knowledge.

Our selection of quotations starts with some general observations on the weather especially the rains, and the observed changes that are occurring.

> The weather has changed too much, there is more heat and more wind, there are crazy rains and it is harmful. Very clumsy rains come, and hail falls. The weather is not good like before, it harms the crops. (M, aged 80, Chorojo)

> Before, the rains were quiet, smoother and very constant. It was not like now, these crazy rains that make rivers everywhere. There is hail as well. (M, aged 72, Chorojo)

> In August, when Antonio [nickname of the Andean fox, Lycalopex culpaeus] goes down [to the river] and howls, it is a sign for a good year, if he howls normally it is a good sign. When it is for a bad year, he does not go down, he only walks on the hilltops, and he howls intermittently. (M, aged 37, Chorojo)

> For the rainy season, there is an important secret to know if it will rain or not, in the stone.... Stones sweat like humans do. So, when it has a couple of flakes, like snow, that year will have a good rain.... In the month of August, you have to watch the stone's temperature.... You have to see its temperature, it attracts ....heat. Then when there is much heat, it will not sweat any more, the heat dries it up. (M, aged 70, Tirani)

Long-term predictions of good or bad years for growing crops are also based on knowledge of nature as it adapts to more chaotic weather patterns:

> I look at the flowers of the muña [Clinopodium bolivianum].... It starts [to flower] in August, just before September which is the time to plant potatoes. So if flowers are well clustered, it is going to be a good year and it flowers abundantly.... If the muña flowers well, I can start sowing tomorrow, and next day, and over next day. (M, aged 42, Chorojo)

When asked what happened if the muña does not flower abundantly, the answer was:

> ... then [you do] not [have to sow] the day after the other. One has to wait for a week between each sowing. Then when the potato flowers, rainfall must come, but sometimes the rain gets lost, so the harvest might be lost, but I have the other plot that flowers later. (M, aged 42, Chorojo)

Finally in our selection, interpretation of climate change becomes linked to broader societal change and the wrath of Gods:

> Before there used to be more respect [for our customs], now there is no respect, this is why the weather has changed as well. If it stays like this, things are going to be worse. (F, aged 57, Chorojo)

> I don't like this change... there is more hail, more rain, unexpected cold or strong heat. This is due to people's bad lives, their infidelities, everybody behaves badly and therefore these things come. For example, in the community some people brought skulls to practise witchcraft. This makes the Pachamama [a regional Goddess] get angry and she does not want to provide us with food any more. (M, aged 24, Chorojo)

> It seems that, as the 'brothers' (people converted to evangelical groups) announced: there will be heat, rains, and landslides; it seems that this is truly happening. This may be the Final Judgment.... Probably in the Bible it is like this, I am thinking about that. (M, aged 37, Chorojo)

Collective lived experiences of groups/communities are therefore made up of complex beliefs and interpretations located in the histories of their specific context. They are not merely an extension of the personal but are shaped by aspects of all kinds of embedded, internalised knowledge regarding climatic and weather conditions.

2.7 Lived Experience and Scale

Scale is of critical importance to both climate change and lived experiences, as summarised in this quotation from Adger et al. (2013: 112): 'Climate change is often portrayed as a global-scale problem: it often does not resonate with the values associated with many traditional, ethnocentric worldviews, and may contribute to antagonism or cognitive dissonance'.

Implicit in this statement are at least two challenges:

1. A climate change analysis is usually at a global scale, offered by scientists who work with large data referring to spatial and temporal scales going back billions of years, whilst that of lived experiences are derived from and defined by people at a human scale probably with reference to themselves and immediate previous generations.
2. Bridging the gap in communication between climate change science and the lived experience of the public.

Firstly, in spite of increasing research attention to links between local and global scales especially from sociologists and related fields, climate change research remains heavy at a global top-down scale and is expected to span the many and varied boundaries that lie within and between global structures and local agency. Yet, scale matters (Wilbanks and Kates 1999; Gibson et al. 2000). Take the pressing problem of food production and distribution, for example. Whilst the assessment of food security and risk is often carried out at global, regional and country scales, for example with the Famines Early Warning Systems (FEWS) http://www.fews.net/ or Food and Agriculture Organisation of the United Nations (FAO) http://www.fao.org/home/en/, there are many variations at local and household scales. These include variations in local landscape and topography which will effect local climatic variations, sometimes contradicting the global or more general-scale climate projections and models. At a country scale there is an enormous variability in the spatial distribution of national Government or international aid distribution, often depending on topographical elements and water availability. For example, in India, investment in the Punjab region of India for new agricultural technology bypasses other neighbouring states such as Bihar (Datta 1992; Fujita 2014). Equally, at a household-scale, prediction of risk and food security is dependent on complex interconnected factors of yield, entitlement, endowment and local agency.

Classifications of climatic zones were produced by the Greek Scholars Parmenides (fifth century BC) and Aristotle (fourth century BC). Fast forward to 1884 and Wladimir Köppen first published his classification, variations of which are still widely used today (Köppen 1936). Typically, these classifications have been presented in the form of maps of the known world. Today's climate modellers also use maps to demonstrate what might be the case under given assumptions at a date in the future. Figure 2.1 is taken from the latest United Nations Intergovernmental Panel on Climate Change Report (IPCC 2013: Chap. 11, Fig. 11.10).

Turning to the second challenge, how does this and other, more complicated, modelling of the future climate relate to lived experience? The answer is, not a lot, for reasons ranging from difficulties in communicating the complexity to the public to the public not wishing to engage, which we discuss in Chap. 7. This issue extends, moreover, beyond the physical science basis as many contemporary modellers attempt to predict the human impacts of climate change in relation to vulnerability, risk and resources such as water, food and agriculture. These human impacts potentially link

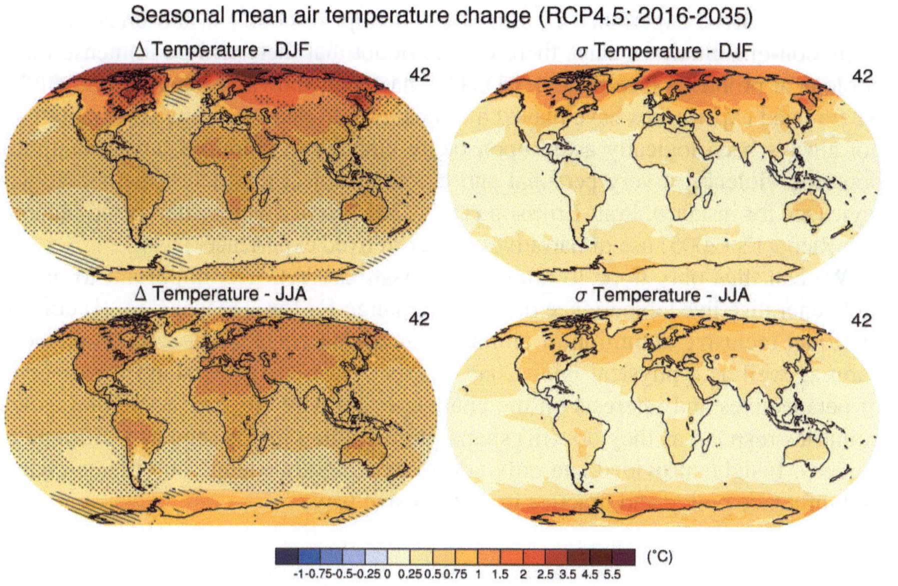

Fig. 2.1 Multi-model ensemble mean of projected changes in December, January and February (*DJF*) and June, July and August (*JJA*) surface air temperature for the period 2016–2035 relative to 1986–2005 under RCP4.5 scenario (*left panels*). RCP4.5 refers to representative concentration pathway 4.5, one of four greenhouse gas atmospheric concentration trajectories that are used by the IPCC. It is the second most conservative of the four trajectories): 2016–2035. The *right panels* show an estimate of the model-estimated internal variability (standard deviation of 20-year means). Hatching in *left-hand panels* indicates areas where projected changes are small compared to the internal variability (i.e., smaller than one standard deviation of estimated internal variability), and stippling indicates regions where the multi-model mean projections deviate significantly from the simulated 1986–2005 period (by at least two standard deviations of internal variability) and where at least 90 % of the models agree on the sign of change. The number of models considered in the analysis is listed in the *top-right* portion of the panels; from each model one ensemble member is used

directly with lived experience, and there is little doubt that such models provide useful and often essential information, but public engagement remains limited.

Climate change problems are multi-scale and, as Cash and Moser (2000) show, such challenges require a linking of science and policy across scales. They thus argue for a recognition of scale and cross-scale dynamics in understanding and addressing global environmental change. This argument can be carried over to a human scale of lived experience in order to communicate more effectively between the more general and bigger models and the local scale of climate change. Again, Chap. 7 explores further these issues.

******

So what is the point of deciphering this maze of lived experience, particularly as unlike 'pure' scientific knowledge we cannot easily offer numbers and statistical analysis of 'lived experience'? Even if we attempt to codify it and create patterns using non-quantitative means, there is little doubt that there will be immense methodological problems (see Sect. 4.5.1). This has been shown in other subject studies where lived experiences have played a critical role in understanding issues, but have not always been logically acted upon being clouded in emotions. One example is domestic violence, a very personal and emotive subject that defies methodological logic (see for instance, Pratt-Eriksson et al's (2014) study which attempts to incorporate Ricouer's (2005) use of narratives to record lived experiences).

We can thus only hope to draw generalisations. Yet, it is important to narrate and generalise lived experience of climate change for three reasons, to: (i) contribute to an inclusive definition of climate change that considers it a social phenomenon as well as a physical one (Chap. 1, Sect. 1.2); (ii) understand the diversity of perspectives and interests on the challenge and why citizens, communities and countries respond as they do; (iii) shape public policy that is seen as legitimate by citizens. Insights provided can only add to existing knowledge on climate change and influence policy and practice in an inclusive way.

References

Adger, W. N., Barnett, J., Brown, K., Marshall, N., & O'Brien, K. (2013). Cultural dimensions of climate change impacts and adaptation. *Nature Climate Change, 3*, 112–117.

Agarwal, B. (1986). *Cold hearth and barren slopes: The woodfuel crisis in the third world.* Riverdale, USA: The Riverdale Company.

Agarwal, B. (2010). *Gender and green governance: The political economy of women within and beyond community forestry*. Oxford: Oxford University Press.

Allen, T., & Thomas, A. (Eds.). (2000). *Poverty and development in the 1990s*. Milton Keynes: Oxford University Press and Open University Press.

Beck, U. (2010). Remapping social inequalities in an age of climate change: For a cosmopolitan renewal of sociology. *Global Networks, 10*, 165–181.

Berger, P. L., & Luckmann, T. (1967). *The social construction of reality: A treatise in the sociology of knowledge*. Harmondsworth: Penguin Books.

Boillat, S., & Berkes, F. (2013). Perception and interpretation of climate change among Quechua farmers of Bolivia: Indigenous knowledge as a resource for adaptive capacity. *Ecology and Society, 18*(4), 21. (Special feature on traditional ecological knowledge and global environmental change: North and south perspectives).

Branisa, B., Klasen, S., Ziegler, M., Drechsler, D., & Jütting, J. (2014). The institutional basis of gender inequality: The social institutions and gender index (SIGI). *Feminist Economics, 20*, 2.

Cash, D., & Moser, S. (2000). Linking global and local scales: Designing dynamic assessment and management processes. *Global Environmental Change, 10*(2), 109–120.

Chowdhury, A. M. R., Bhuyia, A. U., Choudhury, A. Y., & Sen, R. (1993). The Bangladesh cyclone of 1991: Why so many people died. *Disasters, 17*, 291–304. doi:10.1111/j.1467-7717.1993.tb00503.

Corner, A., Markowitz, E., & Pidgeon, N. (2014). Public engagement with climate change: The role of human values. *WIREs Climate Change, 5*(3), 411–422.

Dankleman, I. (2011). Climate change is not gender-neutral: Realities on the ground. *Public hearing on women and climate change*, European Parliament. 11 October 2011. http://www.europarl.europa.eu/document/activities/cont/201110/20111019ATT29752/20111019ATT29752EN.pdf. Accessed 26 February 2015. (This presentation is partly based on Irene, D. (2011). Klimaatverandering heeft een menselijk gezicht. *Streven*, pp. 565–574; Irene, D. (Ed.). (2010). *Gender and climate change: An introduction.* London: Earthscan).

Datta, M. (1992). *Seeds of plenty, seeds of sorrow*. USA: Bullfrog Films.

Drew, G. (2013). Why wouldn't we cry? Love and loss along a river in decline. *Emotion, Space and Society, 6*(2), 25–32.

Frankhauser, S., & McDermott, T. (2014). Understanding the adaptation deficit: Why are poor countries more vulnerable to climate events than rich countries? *Global Environmental Change, 27*(7), 9–18.

Fujita, K. (2014). How agriculture in Bihar lagged behind: Implications for future development. In *Inclusive growth and development in India: Challenges for underdeveloped regions and the underclass* (p. 40).

Gibson, C. C., Ostrom, E., & Ahn, T.-K. (2000). The concept of scale and the human dimensions of global change: A survey. *Ecological Economics, 32*(2), 217.

Giddens, A. (1993). *New rules of sociological method*. California: Stanford University Press.

Ikeda, K. (1995). Gender differences in human loss and vulnerability in natural disasters: A case study from Bangladesh. *Indian Journal of Gender Studies, 2*(2), 171–193.

IPCC Fourth Assessment Synthesis Report. (2007). Climate change. http://www.ipcc.ch/publications_and_data/publications_ipcc_fourth_assessment_report_synthesis_report.htm. Accessed 20 February 2015.

Irshad, S. M. (2014). Caste, asset and disaster recovery: The problems of being asset-less in disaster compensation and recovery. *Journal of Poverty and Social Justice 22(2, 6)*, 137–145(9).

IPCC, Kirtman, B., Power, S. B., Adedoyin, J. A., Boer, G. J., & Bojariu, R. (2013). Near-term climate change: Projections and predictability. In T. F. Stocker, D. Qin, G.-K. Plattner, M. Tignor, S. K. Allen & J. Boschung (Eds.), *Climate change 2013: The physical science basis. Contribution of working group I to the fifth assessment report of the intergovernmental panel on climate change* (pp. 953–1028). Cambridge: Cambridge University Press, United Kingdom and New York, NY, USA. doi:10.1017/CBO9781107415324.023.

Köppen, W. P. (1936). Das geographische system der klimate. In W. Köppen & R. Geiger (Eds.), *Handbuch der Klimato – logie* (pp. 1–44). Berlin: Gebrüder Borntraeger. Cited in Belda, M., Holtanovă, E., & Kalvovă, J. (2014) Climate classification revisited: From Köppen to Trewartha. *Climate Research 59*, 1–13.

McGilvray, D., & Gamburd, M. (2013). *Recovery in Sri Lanka: Ethnic and regional dimensions: Tsunami Recovery in Sri Lanka*. Abingdon: Routledge.

Mendelsohn, R., Dinar, A., & Williams, L. (2006). The distributional impact of climate change on rich and poor countries. *Environment and Development Economics, 11*(2), 159–178. doi:10.1017/S1355770X05002755.

Mills, C. W. (2000). *The sociological imagination* (40th ed.). New York: Oxford University Press. (First published 1959).

Neumayer, E., & Plümper, T. (2007). The gendered nature of natural disasters: The impact of catastrophic events on the gender gap in life expectancy, 1981–2002. *Annals of the Association of American Geographers, 97*(3), 551–566.

Pellow, D. J., & Brulle, R. (2009). Environmental justice. In J. Goodwin & J. Jasper (Eds.), *The social movements reader: Cases and concepts* (2nd ed., pp. 435–440). Bognor Regis: Wiley Publishers.

Pratt-Eriksson, D., Bergbom, I., & Lyckhage, E. (2014). Don't ask don't tell: Battered women living in Sweden encounter with healthcare personnel and their experience of the care given. *International Journal of Qualitative Studies Health Well-being 9(23)*, 166. http://dx.doi.org/10.3402/qhw.v9.23166. Accessed 27 February 2015.

Raleigh, C., Jordon, L., & Salehyan, I. (2008). Assessing the impact of climate change on migration and conflict. *Social Dimensions of Climate Change*. Social Development Department, The World Bank. http://siteresources.worldbank.org/EXTSOCIALDEVELOPMENT/Resources/SDCCWorkingPaperMigrationandConflict.pdf. Accessed 25 February 2015.

Ricouer, P. (2005). *Hermeneutic and the human sciences*, 17th edn. New York: Cambridge University Press. Cited in D. Pratt-Eriksson, I. Bergbom, & E. Lyckhage. (2014). (p. 2).

Seager, J. (2014). Disasters are gendered. What's new? In A. Singh & Z. Zommers (Eds.), *Reducing disaster: Early warning systems for climate change* (pp. 265–282). Berlin: Springer.

Sety, M., James, K., & Breckenridge, J. (2014). Understanding the risk of domestic violence during and post natural disasters: Literature review in L. Roeder (Ed.), *Risks and risk reduction: Humanitarian solutions in the 21st century: Issues of gender and sexual orientation in humanitarian emergencies* (Ch 5, pp. 99–111).

Shiva, V. (1989). *Staying alive: Women, ecology and development*. Zed: New Delhi, London.

Snodgrass, J., Sharma, S. K., & Lacy, M. (2013). Nature reverence does not mean conservation in tribal Rajasthan: Culture, cognition, and personal and collective commitments to the environment. In *Faunal heritage of Rajasthan, India* (pp. 327–350). Berlin: Springer.

The Women's Environment and Development Organization (WEDO). (2008). Gender, climate change and human security lessons from Bangladesh, Ghana and Senegal. http://www.wedo.org/themes/sustainable-development-themes/climatechange/gender-climate-change-human-security. Accessed 27 February 2015.

Wilbanks, T., & Kates, R. (1999). Global change in local places: How scale matters. *Climatic Change, 3*, 601.

Wilcox, A., Harper, S., Edge, V., Landman, K., Houle, K., & Ford, J. (2013). The land enriches the soul: On climatic and environmental change, affect, and emotional health and well-being in Rigolet, Nunatsiavut, Canada. *Emotion, Space and Society, 6*, 14–24.

Wolf, J., Alice, I., & Bell, T. (2013). Values, climate change, and implications for adaptation: Evidence from two communities in Labrador, Canada. *Global Environmental Change, 23*(4), 548–562.

Yoko, S. (2014). Progress or repetition? Gender perspectives in disaster management in Japan. *Disaster Prevention and Management, 23*(2), 98–111.

Zommers, Z., & Singh, A. (Eds.). (2014). *Reducing Disaster: Early Warning Systems for Climate Change*. New York: Springer. doi: 10.1007/978-94-017-8598-3_3.

Chapter 3
Lived Experience and Scientific Knowledge of Climate Change

Abstract Classically, scientific approaches to knowledge, and specifically that of climate change, proceed through systematic collection and investigation of empirical data to establish patterns, and controlled experimentation to demonstrate the truth or otherwise of generalisations, such as 'Human beings are mostly responsible for contemporary climate change'. Scientific approaches also reduce phenomena to their constituent parts, and analyse them independently through subject specialisms, or disciplines, before putting them back together again. Through these controlled processes, science claims to be objective, bringing us closer to reality. Lived experience, on the other hand, proceeds largely through dialectics—the debating of opinions. It is clearly subjective, but nevertheless presents complementary realities. This chapter further argues that (1) Science (natural and social) and lived experiences are each partial versions of reality. The former is limited by the epistemological boundaries that are placed around the subject specialisms that discipline us into certain ways of thinking about phenomena. Lived experience is by definition limited by the practice of everyday life. (2) Science also proceeds through challenge by other scientists of interpretation of empirical results, often starting from different premises. This process is not dissimilar to the dialectics that characterise the process of lived experience. (3) Scientific challenge in relation to climate change is often not neutral or benign. It is seized upon politically to attempt to neutralise the power of scientific evidence in climate change policy debates. (4) The institutionalised perception among scientists themselves that their knowledge is superior, however, leads to significant challenges of working together to generate a more holistic approach to climate change, as illustrated by a personal story of one of the authors.

3.1 Introduction

Chapter 1 started with the lived experiences in the United Kingdom (UK) of flood victims during the unprecedented wave of storms that hit southern England between December 2013 and February 2014. Later in the chapter it compared these stories of relatively affluent people with those of poor people living in

D. Abbott and G. Wilson, *The Lived Experience of Climate Change*,
DOI 10.1007/978-3-319-17945-2_3

the United States and East Africa, thus illustrating the significant role of socio-economic context in mediating lived experience.

The experiences of the UK flood victims did not, however, constitute the only kind of knowledge that was brought to bear on these events. Unsurprisingly, the much revered UK Meteorological Office (affectionately known as 'The Met') also studied them, and arguably it was much more powerful in terms of its influence on policy makers. Here is the essence of its story.

In February 2014 the Met published on its website a lengthy scientific analysis of the extreme weather that had led to the extensive flooding. Its perspective was explicitly 'global' where it linked the British weather of that period to extremely cold weather in North America and wet weather in the West Pacific. It also explicitly addressed the question of whether climate change was a significant contributor. Partly this was in response to the opinions of some flood victims that climate change was indeed the cause (as expressed by two of the Guardian newspaper interviewees—see Chap. 1), partly because environmental groups were vociferously making the same claim, and partly, but most influentially, because the UK Prime Minister David Cameron expressed in the House of Commons that there was probably a link.

The Met's summary scientific explanation of the winter floods in the UK December 2013 to February 2014 is reproduced in Box 3.1.

Box 3.1 A Global Perspective on the Recent Storms and Floods in the UK February 2014 (from the UK Meteorological Office)

This winter the UK has been affected very severely by an exceptional run of winter storms, culminating in serious coastal damage and widespread, persistent flooding. This paper documents the record-breaking weather and flooding, considers the potential drivers and discusses whether climate change contributed to the severity of the weather and its impacts.

This series of winter storms has been exceptional in its duration, and has led to the wettest December to January period in the UK since records began. Heavy rains combined with strong winds and high waves led to widespread flooding and coastal damage, causing significant disruption to individuals, businesses and infrastructure.

The severe weather in the UK coincided with exceptionally cold weather in Canada and the USA. These extreme weather events on both sides of the Atlantic were linked to a persistent pattern of perturbations to the jet stream, over the Pacific Ocean and North America.

The major changes in the Pacific jet stream were driven by a persistent pattern of enhanced rainfall over Indonesia and the tropical West Pacific associated with higher than normal ocean temperatures in that region. The North Atlantic jet stream has also been unusually strong; this can be linked to exceptional wind patterns in the stratosphere with a very intense polar vortex.

As yet, there is no definitive answer on the possible contribution of climate change to the recent storminess, rainfall amounts and the consequent flooding. This is in part due to the highly variable nature of UK weather and climate.

Nevertheless, recent studies have suggested an increase in the intensity of Atlantic storms that take a more southerly track, typical of this winter's extreme weather. There is also an increasing body of evidence that shows that extreme daily rainfall rates are becoming more intense, and that the rate of increase is consistent with what is expected from the fundamental physics of a warming world.

More research is urgently needed to deliver robust detection of changes in storminess and daily/hourly rain rates and this is an area of active research in the Met Office. The attribution of these changes to anthropogenic global warming requires climate models of sufficient resolution to capture storms and their associated rainfall. Such models are now becoming available and should be deployed as soon as possible to provide a solid evidence base for future investments in flood and coastal defences (UK Meteorological Office 2014).

Box 3.1 is a summary of a 29-page report (Ibid.) that documents the frequency and intensity of the storms that brought the flooding, and the behaviour of the jet stream as the immediate physical cause of this extreme weather. Below is information from the full report which throws further light on the analysis.

- It is more accurate to call the jet stream the North Pacific and North Atlantic jet stream as there are others in the southern hemisphere. The North Pacific and North Atlantic jet stream comprises strong westerly winds in the upper atmosphere which are a driver of weather. The weather we receive in the northern hemisphere is largely a function of our position in relation to it and its behaviour.
- Briefly, and simplifying, persistent higher than average rainfall in the tropical regions of the western Pacific Ocean impacted on the jet stream. The higher than average rainfall was due to higher-than-average sea temperatures (causing higher evaporation) in these regions and low atmospheric pressure. These conditions were

compensated on the eastern Pacific (what is confusingly called the western seaboard of North and Central America) by lower-than-average sea temperatures and high pressure. The high pressure pushed the jet stream towards the North Pole in a long wave that then drifted southwards across Canada and the United States bringing extremely cold air with it. Then, as it reached the Atlantic Ocean, it met warmer air and combined with a secondary jet-stream branch, causing the deep depressions of wet and windy conditions that the UK and much of Western Europe experienced.

- Extensive flooding from the sea occurred along the south coast of England, the most dramatic picture being that of a railway line washed away. It could have been much worse, because the impact of strong wave surges was mitigated by improved defences after the major floods in 1953 that wreaked havoc and cost many lives. It was exacerbated, however, by sea level rises over the twentieth century, (for example, 12 cm for the English Channel), this being in turn primarily a function of global warming which has caused seas to expand in volume and receive flows from melting ice.
- The report was reluctant to attribute the extreme weather pattern itself to anthropogenic (human) global warming through greenhouse gas emissions and hence climate change (see Chap. 1, Sect. 1.2 for explanation of these terms and the links between them). This problem of attribution of cause and effect bedevils climate science. While the climate models predict more uncertain and extreme weather, from droughts to floods, from extreme cold to heat waves, it is still impossible to do the opposite using existing computing power; that is, to attribute conclusively specific events when they occur to anthropogenic global warming. The primary factor causing a specific event may simply have been natural variation. The Met report cited Peterson et al. (2013), editors of a special edition of the Bulletin of the American Meteorological Society that sought to explain extreme weather events from a climate perspective. In the 'Conclusions and Epilogue' of this special edition they explained the difficulty through a driving analogy:
  - Adding just a little bit of speed to your highway commute each month can substantially raise the odds that you'll get hurt some day. But if an accident does occur, the primary cause may not be your speed itself: it could be a wet road or a texting driver (quoting UCAR [University Corporation for Atmospheric Research] 2012)

Of course, in this driving analogy, a combination of all three factors—speeding, a wet road and a texting driver—may have been the overall cause of the accident, and separating out the primary cause would be an extremely difficult task. So too, it is very difficult to attribute cause and effect of particular weather events.

Despite its overall reluctance to make the link, the report did offer the possibility, however, acknowledging the evidence that climate change is likely to result in more extreme and variable weather: 'This increase in the frequency/intensity of extreme daily rainfall events, as the planet warms and the atmosphere can hold more water, has been discussed in the literature for a number of years, and robust evidence for this is increasingly seen around the world (UK Meteorological Office 2014: 22, 23)'. It added that the British storms were consistent with both this evidence and the fundamental laws of physics that concern a warming world (i.e. there will be greater evaporation from the seas and therefore more water in the atmosphere). The report also cited evidence (Francis and Vavrus 2012; Petoukhov et al. 2013a) that suggests a link between raised temperature resulting from global warming and a 'locked' jet stream that would in turn produce prolonged weather patterns as happened in the UK December 2013–February 2014 storms.

3.2 Science and Lived Experience: Whose Reality?

The detailed scientific analysis of the Met Office and the reflections of the flood victims in the Guardian newspaper that we examined in Chap. 1 represent two obviously different perspectives on the same event. One narrates and attempts to make sense of the very human experience of the event. It is subjective to the core. The other is a dispassionate analysis that would surely claim to be objective to the core.

There is also no doubt that physical science is viewed as the key knowledge generator for climate change debates and rightly so. It was physical science that brought the issue of anthropogenic global warming to mainstream public attention in the late 1980s. The political figure who first raised the issue as a matter of international importance and concern was the then UK Prime Minister, Margaret Thatcher, in an address to the top British science institution, The Royal Society, in 1988. Margaret Thatcher had, herself, once been a research chemist. In the same year the United Nations Environment Programme (UNEP) and the World Meteorological Office jointly set up the Intergovernmental Panel on Climate Change (IPCC) to improve understanding about global warming. The IPCC has produced periodic reports since and gained a significant reputation. As Blackmore (2009) writes, 'The IPCC is an extraordinary example of international and interdisciplinary collaboration between scientists and other academics across the world. Their efforts have advanced significantly our understanding of how the earth's physical and biological systems, its atmosphere, oceans, land, ice and the living world including ourselves, interact and influence each other'.

Degree of influence and acknowledgements aside for the moment (but see later in this chapter), which of these perspectives—science or lived experience—represents reality? To help answer this question it is necessary to return to the origins of science, its epistemology and method. Although contested, these origins are often attributed to the ancient Greek philosopher Aristotle over 2,300 years ago. Here we lean heavily on a book by the contemporary evolutionary biologist Leroi (2014) which traces meticulously how Aristotle understood and practised science. In comparing Aristotle's methods and philosophy with that of present-day science, Leroi makes the following arguments (page reference numbers to Leroi's book are in brackets):

1. A scientist is someone who seeks, by systematic investigation, to understand experienced reality (39), while 'knowing in the sense of understanding is a requirement for wisdom' (40). Thus, quoting the famed late nineteenth/early twentieth century mathematical biologist d'Arcy Wentworth Thompson, Aristotle turned a 'wealth of natural history' that 'belonged to the farmer, the huntsman and the fisherman' into a science (74).

2. Popular lore is a good place to start, but investigation of the natural world requires expertise, including that of a specific subject (physics, chemistry, biology, etc.) (46). Aristotle, for example, had axioms that derived from popular lore. These axioms about the natural world that he studied were often linked to economics and underpinned by the teleological (Box 3.2) assumption that nature always does something 'because…'—for a reason. Thus, quoting Aristotle directly: 'Like a good housekeeper, nature is not in the habit of throwing away anything from which anything useful can be made…Nature does nothing in vain…What nature takes from one place it adds to another… Nature does not act out of cheapness'. These axioms lead Leroi to conclude that Aristotle 'writes as though nature is living next door and running a taverna' (147, 148).
3. As in modern science, Aristotle looked for patterns in empirical data. Thus Leroi comments: 'Aristotle's appetite is insatiable and his zeal for ordering it tireless… [It is] the empirical rock upon which scientific reasoning stands' (42).
4. Good science is the sense of which causal claims are sound and which are not (123).
5. As Aristotle attempted to practise, the basis of scientific method is *demonstration* of the truth or otherwise of a generalisation, for example, 'human beings are mostly responsible for climate change'. As in this example, such a generalisation concerns universals rather than particulars. Further, Aristotle contends that demonstration is to be distinguished from *dialectics* which concerns debating opinions (124).
6. The aim of natural science is not simply to accept the statements of others, but to investigate the causes that are at work (355).
7. Experimentation is a key component of science. Experiments involve comparison of a deliberately manipulated situation to an un-manipulated control for the purposes of testing a causal hypothesis (362). They are also replicable, in that anybody who follows the original experiment exactly will obtain the same results.

Box 3.2 Teleological Reasoning

This refers to the explanation of something—in the above 'nature'—by the ultimate purpose that it serves rather than by what you think has caused it (a creator or a big bang, etc.).

While Leroi does not address the question directly, he does give an overall sense that science delivers a greater understanding and wisdom than lived experience alone. He makes the point that such understanding invariably starts with the latter (although he does not use the term 'lived experience' explicitly) and associated common sense axioms, but subsequently requires systematic investigation, subject expertise, demonstration and establishing sound causal claims.

We have given some of Aristotle's own axioms above. A basic, probably unspoken, axiom that forms the beginning of climate science investigations might be: 'The temperature of the air that we breathe has no inherent thermostat'. That is, although there is some kind of balance between the energy that the earth receives from the sun and that which it emits into outer space, this balance may be altered by mechanisms that trap more of the energy received, thus reducing the amount that is emitted. One such trapping mechanism involves increasing the concentration of carbon dioxide and other gases in the atmosphere through human activity (see Chap. 1, Sect. 1.2).

In contrast, an underlying axiom of we, the authors, for this book might be to adapt the statement of Karl Marx, to which we referred in Chap. 2 and whose contribution to social science we will describe in Chap. 4, to read:

> Human beings make their lived experiences of climate change, but not in circumstances of their choosing.

We should not, however, take axioms that are derived from popular lore or any other source as static. They might represent a linguistic expression of lived experience, but they will evolve and change along with that experience. As such, questions for scientific investigations will also change. Sometimes an axiom might even be replaced. To take an environmental example, discussed by Hajer (1995: 62–64): in the early industry of nineteenth century Britain, factories were notorious for belching from their chimneys noxious gases and tiny soot particulates into the atmosphere. Was the over-riding concern that of air pollution and associated effect on health? While undoubtedly such a concern existed, it was upstaged by a positive, economics-related axiom (which Hajer calls a 'story-line') in the north of England: 'Where there's muck there's brass' ('muck' being colloquial English for what was belched into the air, 'brass' being colloquial for wealth and ensuing employment). Fast forward now to the late twentieth century and the axiom changed completely to convey a more negative connotation: 'What goes up must come down'. This axiom was the starting point for investigations regarding anthropogenic air pollution, including respiratory environmental health and 'acid rain' that was attacking Europe's forests.

For Aristotle, at least in Leroi's rendition of him, systematic investigation that ensues from axioms involves the empirical work of collecting data, making patterns and generating classifications of what is there. Yes, Aristotle was what we would call today, an empiricist (see Chap. 4). As Leroi (2014: 8) notes: 'Ask Aristotle what fundamentally *exists* … He'd point to a cuttlefish and say—*that*'. [Emphases in original]. He was also, in part a reductionist, for his systematic investigation [of living things] would involve taking them apart and reducing them to their individual bits and pieces to ascertain their functions and mechanisms. Yet, he also indicated that we must then put them together again for it is only when we have done that may we really understand how they work (Ibid: 177). As Leroi notes at the start of his book, 'Aristotle was, if nothing else, a systems man' (Ibid: 6).

Climate science too involves establishing patterns, especially patterns of change. At what rate are carbon dioxide and other 'greenhouse' gases increasing their concentration in the atmosphere? What is the association between increased carbon dioxide in the atmosphere and global warming? To what extent, and where, is sea level rising? How fast and where are glaciers melting? In what ways are atmospheric and ocean circulations changing? In what ways are weather patterns changing in different parts of the world? And so on. This science also, by necessity, refers to the interconnections between these constituent questions and the 'climate system'. Aristotle would undoubtedly have approved.

In a generalised sense, the processes parallel the ways in which we put our lived experiences to use. Pattern-making seems to be inherent in human sense-making of our lives. We make a pattern of what has and what is happening to us. We establish patterns with lived experiences of others, either through direct engagement or through communication media such as reading, sound and visuals. We may even be inclined to reduce our lives to simple, or even single, causal mechanisms, including those associated with climate change. For example, before the UK floods 2013–2014, many of those who were affected loved their homes, but now, in the words of one victim, cannot 'feel so sentimentally attached to it' solely because of that event.

One should not, however, make too much of these parallels because it is *not* part of our argument that lived experience and scientific investigation are at root the same. Rather, both represent fundamentally different attributes of our ability to make sense through reflection and engagement. Let us, therefore, note the parallels and then move on to Aristotle's claim that subject expertise is necessary for greater understanding. This is where science and lived experience certainly diverge. Science thrives on subject expertise for deepening understanding of a particular component part of a system. Thus physicists investigate fundamental mechanisms underlying climate change, including the energy balance and the energy-absorbent capacities of carbon dioxide and other atmospheric greenhouse gases. Meteorologists explain changing weather patterns in terms of atmospheric and ocean circulations, and their interconnections across the world. Economists explain probable impacts on regional economies in different parts of the world and the implications of their interconnections for the global economy. Sociologists explain the probable differentiated human impacts on societies and social psychologists explain why it is so difficult to get through to people about climate change, despite the science being laden with doom warnings.

There is no parallel to any of this in lived experience, although we might be influenced from time to time by popular renditions of different scientific assessments. Undoubtedly, subject expertise provides different angles and perspectives, different understandings if you like, of climate change. We prefer 'different' or 'diverse' understandings as more neutral terms than 'deepen understanding' which has value-laden undertones, and where we must ask the question, 'deepen for whom'.

While we acknowledge the contribution of diverse subject understandings, we also recognise two issues. First, subject investigations are constrained by the questions they seek to answer which in turn derive from the conceptual frameworks and theories that define the particular subject matter. It is no wonder that different subject areas are called disciplines (Chap. 4), because that is what they do to our thinking. For example, reference points for physicists on climate are the laws of thermodynamics and modes of energy transfer. For classical economists, they are markets, innovation and cost–benefit analysis. For sociologists they are social power and inequality among social groups differentiated by race, gender and class (Chap. 2). Application of these frameworks certainly provides important understandings, but they also constrain knowledge in that they put boundaries around what questions may be addressed and the means of discussing them.

The second issue is the danger that subject specialisation may easily take on a life of its own. Partly this relates to the first issue, because the epistemological boundaries of each subject risk becoming impermeable to other subjects. It also raises the danger, however, that the subject becomes completely divorced from the original axioms that gave rise to investigation, axioms that are rooted in lived experience. In other words, there is a disconnect between the starting and end points of the investigation. The subject is divorced both from other subjects and from lived experience. This disconnect, we argue, degrades our *systemic* knowledge of climate change. (See Chap. 2, Box 2.3 for a further explanation of systems and systemic explanations.)

Experimentation is a further key aspect of modern science where there are no direct parallels with lived experience. It is true that this book has, in part, conceptualised lived experience as a string of action-research experiments, where we try something different to gain an improvement, continue with it, adapt it or discard it depending on the results. This common sense use, however, bears little resemblance to the carefully controlled, targeted experiments that Leroi defines above. For Leroi (2014: 368): 'Data trawls and pattern analysis give you models; targeted experiments tell you whether or not they are true, Many [modern] scientists use both'. Figure 3.1 shows a modern experiment, whose aim is to investigate the possibility of increasing tomato yield in Morocco using less water (and less pesticide and fertiliser).

We should note in passing that Aristotle did not do experiments in the modern sense; also in French that the same word—expérience—is used to mean both 'experience' and 'experiment'. Of direct relevance to this chapter, however, is the challenge of experimentation for climate science that can be traced back to 1859. As recounted by Hulme (2009: 42–47), in that year and for several thereafter, the Irish scientist John Tyndall provided the experimental verification for the proposition over 30 years earlier of the French physicist Jean-Baptiste Joseph Fourier that the constituent gases of the atmosphere trap some of the incoming solar energy, what later became known as the greenhouse effect. Tyndall established experimentally that molecules of water vapour, carbon dioxide, nitrous oxide and ozone each had particular absorptive properties for radiant heat. Then, in 1895, the Swedish Physicist, Svante August Arrhenius established the link between carbon dioxide concentration in the atmosphere and variations in the mean surface air temperature of the world. He performed this feat through calculations (done by hand) based on the earlier discoveries of Fourier and Tyndall, and in that sense the link was still theoretical, rather than demonstrated empirically.

Since these nineteenth century discoveries, climate science has been driven forward less by experimentation, and more by measurement and correlation between greenhouse gas compositions in the atmosphere and temperature. In other words, patterns

Fig. 3.1 The 'fertigation' experiment to maintain or improve tomato yield using less water, pesticide and fertiliser in Morocco. Tomatoes are an important export crop in Morocco and are bought by large European supermarkets. This experiment uses 'smart' computer-controlled irrigation to water the plants at optimum times during the day (instead of constantly). 'Leaky' pipes in the tomato troughs are connected to a 'fertigation system' that mixes liquid fertilisers with water, the mix being varied at different stages of plant development to give optimum growth. Pesticide use is lowered by improved physical conditions for growing the crop. The early results are promising, reducing the water wastage of conventional irrigation and the import of fertiliser from Spain. The project is funded by the European Union under the heading, 'Sustainable Use of Irrigation Water in the Mediterranean Region' (*Photograph* Oliver Tomlinson 2014. *Location* Institut Agronomique et Vétérinaire Hassan II, Rabat, Morocco)

have been established from data, just as Aristotle established patterns. From them, models of the changing climate have been (and are still in the process of being) created.

The enduring problem of these data trawls and the patterns that have resulted in ever sophisticated models of the current climate and future projections is that they are always susceptible to alternative interpretations and, indeed, the generation of alternative models. This is essentially a problem of attribution, that is, what is attributable to what? In a controlled, targeted experiment one controls everything apart from the variable whose influence one is trying to establish. One exercises such control in a laboratory but the world's climate system can't be replicated in a laboratory. Measurements have to taken in situ, that is, in the existing world where one is no longer in control of the variables and their influences on each other. It means that one can never be completely sure and it is commonplace for others to question or even dismiss findings.

Thus to continue the story of science and climate change, the first person to provide, in Hulme's (2009: 50) words, a systematic link between 'the physical theory of carbon dioxide and the greenhouse effect, the rising concentration of carbon dioxide in the atmosphere and the increase in world temperature' was the British engineer Guy Stewart Callendar in 1938. His audience was largely sceptical—from meteorological professionals, and others who questioned the reliability of his data and his calculations.

Other scientists, however, continued the studies. In the late 1950s American scientist Charles David Keeling conducted experiments to measure systematically carbon dioxide concentrations in the atmosphere over time (Ibid: 54–56). In 1975, Syukuro Manabe established in the United States the first computer model for establishing climate sensitivity to carbon dioxide concentrations.

Nevertheless, the concern that climate change is induced by anthropogenic activity remained largely a preserve of natural scientists until the late 1980s. Then, in 1987, drawing on mounting evidence which showed that climate change in the past could have occurred much more rapidly than the slow moving oscillations over thousands of years that had previously been the consensus, American oceanographer, Wallace S. Broeker, argued in the influential journal 'Nature' that human activity in the future might trigger abrupt changes in aspects of the Earth's climate (Ibid: 59–60). Broeker was not the only scientist making the claim, but from this moment on we can identify climate change as a matter of public concern, often dovetailing with, and initially being subsumed within, broader environmental concerns. The following year—1988—Margaret Thatcher made her famous speech to the Royal Society in London while the United Nations Environment Programme and the World Meteorological Organisation established the Intergovernmental Panel on Climate Change (IPCC) which has reported regularly since (see above).

There are some who continue to deny global warming as a long-term trend, arguing from a statistical perspective that the time frames are too short to establish a definite pattern and that recent evidence is just part of natural variation. There are others who might find evidence of global warming but argue that it's impossible to separate out with full confidence the proportional impacts of natural (for example, changes in sunspot activity) and anthropogenic sources. The IPCC reports, meanwhile, state with increasing confidence the overwhelming scientific evidence that global warming is occurring and that it is most likely due to human activity. While the inference is greater than it once was, Leroi's (2014: 129–130) general comments about data that are gained from study of the 'natural world' still hold. They are 'incomplete, results are tentative and inferential gaps yawn at every turn'. This is why the IPCC qualifies its findings in terms of level of uncertainty, degree of confidence in validity and likelihood (see Chap. 6, Box 6.1).

All of this brings to mind Leroi's citation of one of his colleagues that natural experiments are not experiments at all, because the only variables that should differ between control and treatment are those manipulated by the experimenter. But when you rely on nature to do your manipulations you can never be sure what she's meddled with (Ibid: 128–130).

The models of how the climate behaves and will behave, depend on [probabilistic] answers to 'what if questions', for example, 'What is the most probable rise in global temperature on current trends?' What is the most probable rise in global temperature if carbon dioxide emissions accelerate?' 'What is the most probable rise in global temperature if international action is agreed to reduce global carbon dioxide emissions by X%?' These questions then lead to equally probabilistic further questions about impacts. 'What if the global temperature rises by 2 °C?… What if it rises by 5 °C?'

According to Leroi, cited above, the basis of Aristotle's scientific method was 'demonstration'. In climate science, demonstration depends largely on the construction of models. Modern computing power enables them to handle huge amounts of data, and they have become very sophisticated, with increasing resolution to identify local climate change patterns. They remain, however, models—representations of the climate and possible future climates for demonstration purposes. To reinforce the point, they generate probable patterns of climate but they are not rigorous proofs.

Dealing with probabilities helps deal with the question of soundness of causal claims in climate science. Probability contains an in-built 'get-out' clause, because, whatever the odds, it is not beyond the bounds of possibility that the expected causal claim will not be borne out. Further, if it is not borne out, the science then progresses by asking 'why'? For example, under a business-as-usual scenario, it is predicted that global sea levels will *probably* rise in more than about 95 % of the ocean area by the end of the twenty-first century (IPCC 2013: 26). They might, however, rise by a significantly different measure, at which point the science is asked to suggest explanations for the discrepancy, just as senior politicians asked that the latest (2013) IPCC report provided an explanation for an observed 'pause' in global warming over the past few years. In this sense, being 'inaccurate' about a probabilistic prediction is simply part of the process of increasing understanding.

Based on demonstration, it was through the overall quality of his scientific methods that Aristotle sought *soundness* of his findings. Modern science is also concerned with the quality of its scientific method to achieve soundness, but that is not all. Thus, Hulme (2009: 75) argues forcefully that science 'thrives on disagreement. Science can only function through questioning and challenge'. Through this process, involving a peer community, a causal claim is rarely validated without qualification, and such validation will always be provisional as fresh evidence and insights are brought to bear.

A good example of 'science in progress' through questioning and challenge may be found in a study cited by the UK Met Office in its aforementioned scientific report of the 2013–2014 floods. As noted in Sect. 3.1, the report cites recent studies concerning a possible link between extreme weather events and locking the jet stream into particular patterns. One of these studies (Petoukhov et al. 2013a) identified a physical mechanism to understand the exceptional nature of many recent extreme events associated with persistent large-scale weather patterns. They also suggested that the generation of these large-scale patterns might be affected by processes associated with global warming. Their paper was published in the Proceedings of the National Academy of Science (PNAS) of the United States of America and sparked a letter (Screen and Simmonds 2013) that was published less than three months later in the same journal. The letter stated that their (i.e. Screen and Simmonds') analysis, including use of the same data source as the original paper, showed no statistically significant trends that would be consistent with the proposed mechanism. Petoukhov et al. (2013b) responded in the same issue by stating that their paper concerned the 'physics, not statistics, of extreme events'. In other words, the challenge of Screen and Simmonds was based on a different epistemological framing of the study—statistics, rather than physics. As Petoukhov et al. (Ibid) further pointed out, the two studies appear to be complementary, rather than in opposition. They also both involved pattern analysis in climatic trends, but with different aims. For Petoukhov et al. it is was to propose a physical mechanism that could explain weather extremes; for Screen and Simmonds it was to ascertain whether such a mechanism was consistent with statistical trends.

For many natural scientists, establishing physical mechanisms is at the core of modern science. Thus, once one has been proposed, one should expect a debate about its soundness. In this example, Petoukhov et al. published their findings on establishing a physical mechanism that would explain a possible link between extreme events and large-scale weather patterns that are influenced by global warming. Screen and Simmonds provided a challenge through a statistical trend analysis. Petoukhov et al. responded, clarifying 'some important issues', pointing out the different disciplinary frames of analysis, and the further study that is required.

We should not, therefore, ever expect the science to be 100 % settled.

The above example, given to illustrate Hulme's argument about how science progresses, appears to be little different from the process of 'dialectics'—debating opinions—which Aristotle distinguished from scientific demonstration. Agreed, most scientists would declare that they are debating factual questions and interpretation of results. Even so, this is a process of constructing reality within the confines of their current knowledge. We would further argue that it is also a process that takes place within a social context. Scientists are not, therefore, 'uncontaminated' by subjectivity and social interference. While discovery is part of the process, and a trigger for debate it is not an end point. The discovery is not self-evident—'out there'—lying beyond conflicting human interpretation.

In climate science, the whole IPCC process of sifting through and analysing the scientific outputs on climate change is 'dialectics' writ large (Box 3.3). The process of lived experience too involves 'dialectics' through, as we discuss further with respect to the philosophy of Jürgen Habermas in Chap. 8, personal processes of reflection (debate and challenge within ourselves) and engagement (debate and challenge with others). Again, we do not wish to over-state similarity, but at fundamental human levels there are parallels between scientific processes and our processes of lived experience. It is interesting too that the IPCC is attempting to bring perspectives from different disciplines together via its 'integrated assessments' which first appeared in its 2007 report (see Chap. 8). Integrated assessments require dialectics across disciplines rather than within them. This might appear a daunting task, although the British philosopher Midgley (2014: 36) argues that conflicts of opposing views may be reconciled via dialectics. She adds a condition, however: acceptance of each is required for reconciliation.

Box 3.3 The IPCC and its process

In its own words (http://www.ipcc.ch/organization/organization.shtml):

The IPCC is a scientific body under the auspices of the United Nations (UN). It reviews and assesses the most recent scientific, technical and socio-economic information produced worldwide relevant to the understanding of climate change. It does not conduct any research nor does it monitor climate related data or parameters.

Thousands of scientists from all over the world contribute to the work of the IPCC on a voluntary basis. Review is an essential part of the IPCC process, to ensure an objective and complete assessment of current information. IPCC aims to reflect a range of views and expertise. The Secretariat coordinates all the IPCC work and liaises with Governments. It is supported by WMO and UNEP and hosted at WMO (World Meteorological Organisation) headquarters in Geneva.

The IPCC is an intergovernmental body. It is open to all member countries of the United Nations (UN) and WMO. Currently 195 countries are members of the IPCC. Governments participate in the review process and the plenary

Sessions, where main decisions about the IPCC work programme are taken and reports are accepted, adopted and approved. The IPCC Bureau Members, including the Chair, are also elected during the plenary Sessions.

Because of its scientific and intergovernmental nature, the IPCC embodies a unique opportunity to provide rigorous and balanced scientific information to decision makers. By endorsing the IPCC reports, governments acknowledge the authority of their scientific content. The work of the organization is therefore policy-relevant and yet policy-neutral, never policy-prescriptive.

This brings us to the notion of "post-normal science", which challenges the ability of conventional modern science to deal with complex real-world problems, of which climate change is a prime example. The argument here is that inquiry into such problems through reduction into constituent parts, standard scientific analysis involving controlled experimentation and abstract theory building reveals at best only a partial understanding of the whole. Instead, a complex systems approach is advocated, which recognises that any "system" is itself an intellectual construct and that human beings are embedded in their own systems. In other words, we decide what is included in this system and what is excluded, the interconnections of the parts and the kinds of questions that it is designed to answer. In this way, human subjectivity (and therefore partiality) becomes a factor in the knowledge generation project, although this need not be seen as a limitation. Thus, post-normal science scholars consider that awareness of human subjectivity may enrich the analysis (Funtowicz and Ravetz 2013). It is a view that allows for a re-connection of science with lived experience. The latter is nothing but human subjectivity at work.

To return now to our starting question for this section, which then is closer to reality, science or lived experience? In the final analysis we have to accept that reality might be a vision, but in practice it too is a process that is always in the making, and that at any given time there are multiple realities. 'Sense-making' seems to be a more helpful term that describes this process and it applies to both lived experience and science. Thus, lived experience concerns the ongoing—and therefore provisional—sense that we make of our lives through the accumulated experience of living those lives. To return, however, to the basic axiom of this book, we make and live our lives in circumstances not of our choosing.

It is in relation to this matter of our 'circumstances' that science makes its grand entry. Science stems from the impulses to understand, that is to make sense of, circumstances, dividing them into three broad categories. The social circumstances are the province of the social sciences. The physical circumstances are addressed by the physical sciences, and the living world around us consumes the biological or life sciences. Such sense/understanding is also provisional, proceeding through dialectics as much as demonstration.

Unlike the separate sciences, lived experience appears by definition to be holistic, although that might be disputed. We are, after all, often inclined to attribute cause and effect to some aspect of our lives in a reductionist way—it is all because of X that Y has happened. The sciences meanwhile are driven by further and further inquiry as previous explanations are found wanting. This in turn requires further fragmentation within the broad categories. For example, and the list does not claim to be exhaustive: economics, sociology, political science and anthropology within the social sciences; physics, chemistry and geology within the physical sciences; and zoology, botany, biology and cognitive psychology within the life sciences.

We have also noted how these largely separate areas of scientific inquiry discipline us into certain ways of thinking about problems because of the epistemological boundaries that the core concepts of each set. The very fragmentation, therefore, results in only partial representations of reality that are limited by their epistemological boundaries that establish what may be known. As Midgley (2014: 34) puts it: 'It is not possible for any single department of thought to expect to handle large questions on its own'. We argue in Chap. 7 that lived experiences also have limiting epistemological boundaries, limited by the lives that we lead. Thus neither scientific nor lived experiential representation by itself has an exclusive hotline to reality. Both are incomplete. Following Midgley (Ibid: 142) again, we argue that the two perspectives, while being different in many respects, are not necessarily mutually contradictory and may sit quite comfortably side by side and may indeed combine to provide an enhanced grip on reality. To take another analogy from Midgley's book, 'Are you an illusion?' (Ibid.), there's little point telling somebody who describes a table as 'solid' that actually it is mainly empty space, comprised of atoms and molecules. Likewise, there's little point explaining to a flood victim that actually one can't attribute this specific extreme weather event scientifically to climate change, although that is a possibility. As British nature writer, Mabey (2013: 80), points out, how we live with the weather is 'our daily, intimately experienced embodiment of climate'.

We add to Midgley's analysis of potential complementarity by noting that the difference between science and lived experience is not always as clear-cut as it appears. There are similarities. A more fruitful rendition would be to envisage them as being at opposite ends of a spectrum with shades in between.

Having said that, we must also recognise that all forms of human knowledge exist in social contexts, both in their production and the uses to which they are put. 'Knowledge itself is power', famously wrote the British philosopher and lawyer, Francis Bacon, in 1597. He was right in the sense that knowledge is one source (among others) of social power, but only partially. Knowledge is not undifferentiated, there are many forms of knowledge and some are more powerful than others (see Chap. 11). If we take degree of influence on decision makers to be a good proxy for power, as noted earlier in this chapter scientific knowledge holds great sway, especially in debates about climate change and responding to it. In that sense, politicians and other decision makers seek to derive the truth of their arguments from science. For them it represents the reality of the challenge.

Or so they indicate. The challenges mount when we recognise that there is no such thing as a single, unified science.

3.3 Diverse Scientific Perspectives and the Political Exploitation of Diversity

There exist many differences of scientific perspective on climate change. While none are perhaps quite as fundamentally marked as the deep epistemological divide between 'subjective' experience and 'objective science', they nevertheless present a deep challenge to a unified knowledge on the subject. When we examine science, the first point to note is the obvious one that there are many, not just one, sciences each with different frames for analysing the same event, frames that are bounded by their respective disciplines. Often they may live quite happily side by side, for example an economics analysis of flood impact alongside a meteorological explanation of its causes.

Such diversity, however, does raise the issue that some sciences might be more powerful in their influence on political policy making than others. We return to this subject in Chaps. 9–11. Here we note that, of the sciences, the physical sciences have made the running over climate change. In the previous sections we traced the early breakthroughs of Fourier (1824), Tyndall (1859), Arrhenius (1895) and Callendar (1938), followed by Keeling (1950), Manabe (1975) and Broeker (1987), all of which invoked physics and/or physical chemistry within meteorology.

It was not until this century that the social sciences seriously started to find a slot within climate change debates, as the politics surrounding intervention and the socio-economic impacts have consolidated as fields of study. Thus, significant sociological and anthropological work has now been conducted concerning local impacts, threats to livelihoods and ways of life, uneven impacts across different social groups and different parts of the world, as cited throughout this book (see also, for example Toulmin 2009). The political science of climate change has made complementary inroads (for example, Breitmeier and Otto 2012; Giddens 2007). In terms of influence with politicians and other policy makers, however, it comes as no surprise that economics has been first and foremost among the social sciences. The year 2006 proved to be seminal for this discipline, with the publication of the internationally influential review that was commissioned by the UK government under the leadership of Stern (2006). This review, titled 'The economics of climate change', used a classical economic analysis to conclude that the cost of doing nothing to mitigate climate change far outweighed the costs of doing what was needed now.

The Stern Review certainly did shift the ground, judging by media headlines, for example the Washington Post's 'Warming called [a] threat to [the] global economy' (31 October 2006—cited in Hulme 2009: 125). The discourse did not stop there, however, and in recent years a more positive tone has evolved. For example, in 2010 Stern's later book was published (Stern 2010), with the title 'A blueprint for a safer planet' and a sub-title 'How we can save the world and create prosperity'. Thus, rather than equating failure to act on climate change with global economic doom, which was the impact of his 2006 Review, the ground has shifted further to assert that acting on climate change will create a new kind of economic growth which is predicated on developing green technologies, the skills needed to operate them, and hence millions of jobs. This 'win–win' discourse can be seen in numerous policy proposals. For example, the European Union Roadmap for moving to a low carbon economy by 2050 states, 'This will require targeted vocational training of the existing work force towards "green-collar" job opportunities, addressing emerging skills bottlenecks and fostering these skills in education systems' (European Commission 2011: 12).

What is also interesting about the Stern Review and the later book, however, is the way that both aligned with the prevailing physical science. In fact, the starting point on which the analysis was built was the increasing carbon dioxide emissions linked to global warming. Stern then looked at the likely human impacts, and put economic values on them, together with values on the costs of mitigation, that is stabilising global carbon dioxide emissions. The approach, therefore, was to draw on the physical science to calculate and compare the costs and benefits of doing nothing or taking action, although Stern actually denied that his was a classical cost–benefit analysis.

Diversity among the sciences, formidable alliances between the physical sciences and economics, and 'speaking to political power' form one level of analysis. Another level concerns the disputes within disciplines. We are probably all familiar with the discordant voices of the climate change sceptics, to which we referred briefly in the previous section. Some scientists who accept anthropogenic warming have also suggested that we are addressing, and putting money into addressing, the wrong problem. Thus the physicist, Freeman Dyson, wrote a short essay in 1995 suggesting that we should be examining the productive potential of increased carbon dioxide in the atmosphere to increase plant growth and hence be a valuable weapon for improving global food supply (Dyson 1995: 492–494). Simply put, this argument is based on the principle that plants absorb carbon dioxide through the process of photosynthesis, which is the basis of plant growth and health.

The large majority of physical scientists, however, subscribe to the view of anthropogenic global warming and the threats that this poses to climate systems over the world. Nevertheless, even within this group there are arguments over the rate of increase of carbon dioxide concentration and of temperature rise.

The Stern Review also produced discordant voices among economists. These centred mainly on the value he placed on the cost to future generations of climate change. 'Discounting the future' to a present value, as it is called, is a standard technique in cost–benefit analysis where costs and benefits will extend several or more years into the future. Basically, the assumption is that the costs and benefits will have a declining present value into the future. This declining present value is expressed as a percentage—the discount rate. The choice of discount rate potentially has a large impact on the overall cost–benefit calculation, with a high discount rate lowering significantly the present value of benefits of mitigation action now and a low discount rate having less impact.

The Stern Review assumed a deliberately low discount rate, and hence a high value on benefits, on the basis that we should value the lives of future generations equally to our present lives. Other economists, however, have contended that his value (1.4 %) is far too low and much lower than the conventional range for discount rate of 3–6 % (for example, Nordhaus cited in Hulme 2009: 127–128; Stern 2010: 91–92, 150).

More radical critiques, meanwhile, centred on the choice of a single economic value—gross domestic product—to measure human welfare in the Stern Review, and the danger of not investing to meet other global challenges (such as poverty and global health) because of the priority given to climate change. Chapter 2 has explored these issues, drawing especially on developing country examples.

In the previous section we presented the debate over long wave patterns of the jet stream as part and parcel of the way that science proceeds through questioning and challenge. One might argue the same of the debates over global warming and carbon dioxide concentrations, and the economic debates that surrounded the Stern Review. Such a view taken literally, however, would be naïve. In the hands of politicians, scientific findings are rarely innocent. They are seized upon and used by politicians and whoever is lobbying them to support whatever case they are making (Nelkin 1977). Thus, the Stern Review was commissioned by the UK Government to make the crucial link between the economy and climate change impacts. Its conclusion that climate change would cost the earth dear, and that these costs would far outweigh the cost of taking mitigation and adaptation measures now, were seized upon by the then UK Prime Minister Tony Blair, who proclaimed of the final report: 'It proves [that] tackling climate change is the pro-growth strategy' (cited in Hulme 2009: 125).

If scientific evidence from whatever source becomes a political tool to bolster a case, the situation becomes murkier where there is scientific disagreement. This opens up a space for rival lobbyists to cancel out the science of each other and to disagree on what, if anything, needs to be done. Thus we find that the minority of climate change sceptics mount a challenge to the majority scientific consensus through attempting to discredit the evidence of anthropogenic global warming with rival evidence and the scientists who do subscribe to the consensus. As Nelkin (1977: 189–205) noted many years ago, it matters not that the challenge may or may not be scientifically sound. What matters in political terms is that it has been made and constitutes a challenge to action.

This is the realpolitik of questioning and challenge of scientific evidence at any given point of time. A well-documented environmental (but not climate change) example is provided by Howard (2011), who analyses the doomed attempts to phase out the production of chlorine-based chemicals in North America. The proposals were made in the early 1990s by the Non-Governmental Organisation (NGO) Greenpeace and the 'International Joint Commission' (IJC)—a 'quasi-governmental institution advising the United States and Canadian Governments on border issues'—on the grounds of known harmful effects of many of these chemicals and potentially harmful effects of the rest. Initially, they garnered a positive response, including from senior political figures in both the United States and Canadian governments. The industry that produces these chemicals which have a wide range of uses is, however, huge. Soon an array of scientists and other experts from the industry and elsewhere were mobilised against the science and expertise of Greenpeace and the IJC. Opposing scientists were also certainly not immune from making distinctly unscientific comments. For example, one biologist warned: 'A world without chlorine would resemble a science fiction scene of the worst kind'. By 1993, the proposals had slipped off the political agenda.

These examples illustrate the hollowness of the basic claim of science, still made today, to strive for an objective, and hence 'disinterested', universal truth. Scientists are rarely disinterested and their consequent subjectivity is not necessarily benign. The 'truth' is established in the end by which sciences and scientists speak to power. Thus, the industry scientists and other experts who aligned against phasing out chlorine-based chemicals spoke to corporate power. On climate change, the Stern Review on the economics of climate change combined with the science of global warming speak to the UK government that commissioned it, and to other rich-country governments. That has not, however, been the end of the story where the infamous theft of emails from members of the Climate Research Unit at the British University of East Anglia in 2009 has turned out to be just one significant subsequent event (see Chap. 4 for a fuller exposition of this episode and the claimed objectivity of science). These emails purportedly showed evasion in requests for information and pejorative comments against those who were questioning the science of global warming. They were like manna from heaven to climate change sceptics attempting then, and now, to discredit and neutralise that science on which the Stern Review was based.

3.4 Working with Scientists

The norm of science, especially physical science, is that it is objective and universal. However corrupted it may be in practice, it is because of this norm that science is able to speak powerfully to political power in climate change debates. Midgley (2014: 26–28) argues further that all the other sciences try to be like physics because of a strongly held belief that only this discipline represents reality. Thus, the technical language of the physical sciences is deemed superior, including by politicians and policy makers, because it brings us closer to reality. It is a reductive story about reality, but a powerful one.

Whether corrupted or not, we cannot deny this norm. Nor can we fail to acknowledge the enormous contribution that physical science has made towards highlighting and understanding anthropogenic climate change.

Physical scientists above all institutionalise the norm and their institutions reinforce it in their daily professional lives. We have argued in this chapter that these institutionalising practices tend to induce blind spots with respect to other forms of knowledge.

The realpolitik is that we cannot easily change the norm and the mindsets it produces. This is not to say that it should not be challenged by those who are not physical scientists. Such challenge, however, requires an everyday presence, and that itself is challenging. What is it like to work among physical scientists and their mindsets? Section 3.4.1 tells the story from one of the book authors who has to do just that.

3.4.1 Working with Physical Scientists: A Social Science Story

This is very much a personal, subjective story of working with physical scientists and their mind-sets and I hasten to say that my script may not be that of others. My own experience of working with physical scientists is that this is challenging, and at times a stressful encounter where people are often unable to, or do not want to, step beyond known boundaries.

First, I should point out that I am immersed in the qualitative methodology that informs much, but by no means all, social science which inevitably affects my own standpoint on the matters I describe below. Second, neither social science nor physical science is a homogeneous category and both have several disciplinary strands. For instance, I am a development geographer which means that I already cross two disciplinary strands—that of development studies and human (or social) geography. Such complexities are therefore quite demanding for both groups (social and physical scientists) as they come together in a multidisciplinary effort—which at the same time maintains an essential separation of disciplines throughout the collaboration, making it easy for each to return to their own boundaried comfort zones and onto the next project. Interdisiplinarity on the other hand links together an integrated, merged approach, discovering synergies between disciplines and crossing boundaries with the potential of creating new insights towards a holistic approach to knowledge. Both terms, not always differentiated and often interchangeably used, involve team work and are increasingly advocated by university teaching and research agendas.

I have worked with interdisciplinarity in at least two differing set-ups with mixed teams. One involved working with a team of European academic and practitioner collaborators in producing Masters MSc online modules and resources for a European Union teaching project on climate change, 'The lived experience of climate change: e-learning and virtual mobility' (LECHe—see Chap. 6, Box 6.1). Another occasion was an attempt to incorporate the modules produced under the LECHe project to develop a new online MSC on climate change within the Geography department at the institution that I work for. Here I go through my experiences of both projects as shaped by my own positionality as a development geographer and a social scientist amongst a heavy physical science disciplinary representation. At the same time, the fact that I was the singular person of colour amongst a totally white team in both cases, and as a woman within a mostly male team, cannot be bypassed as these characteristics shape my thought processes and perspective somewhat differently from those of the dominant science paradigm.

My very first encounter with the LECHe European team was one based at the imposing ancient Leuven University in Belgium. The team represented several European institutes all of whom were very nice people, some in very senior positions in their universities. Yet I did not feel comfortable in the first few encounters with the team. This was partly because we all came from differing disciplinary backgrounds, the intricacies of which I was trying to grasp, but mostly because unlike UK Universities where there is a visible Black British presence (faulty as this may be) in the higher echelons, this did not appear to be the case within continental universities. I felt that the team was trying hard to grasp my position as a researcher and an academic, my background, my discipline and my status within a British University. None the less, as time went on over the two years that the project lasted, as we came to know each other better and my contribution to the project became regarded as sound, I felt much more at ease and almost felt I had paved a path for other people of colour and disciplines to work together. I did feel respected, but along with other Black women in universities generally, I had to prove myself first to gain that respect, otherwise more naturally given to other colleagues at the same professional position as myself. Nevertheless, I made new friends and under the guidance of a strong leader, the end of the project was deemed a success.

My second attempt at an interdisciplinary project was when I was tasked by my line manager to lead the development of a new MSc Lived Experience of Climate Change programme at my home institution. The team consisted of colleagues from my department who have known me for a very long time, but some resentment was apparent in that as I am a social scientist, and a senior academic (Professor), the 'scientists' felt they had superior knowledge of climate change and that I lacked authority both in my senior rank and as a social scientist. What followed was massive disruption, caused by one colleague in particular who felt threatened by the process and led other newer and younger members of staff into doing the same. In the end, the idea was abandoned causing much stress and ill feeling within the department.

Getting people to buy into interdisplinary agendas is therefore the most difficult thing to achieve for at least two reasons: (i) the notion of 'bad' science, and (ii) the personal, often unspoken agendas where interdisiplinarity is regarded as a threat. At a general level, there is a camp of those scientists who have open minds, ready to explore the new, and at the other end of the spectrum are scientists who resist with all their might—attempting to fob off social scientists as creating 'bad' science, rather than seeing them as representing strong disciplines in their own right that provide an opportunity to create critical insights. Rejection on the basis of 'bad science' is easy as it does not rely on opening up minds to other fields of knowledge and moving out of personal comfort zones. For instance, in reviewing this book proposal, one reviewer wrote:

> I strongly disagree with the implicit assumption that scientific understanding is influenced by 'lived experience' by individual scientists. I would claim that these are bad scientists. I would by now say that for instance the so-called sceptics are this type of bad scientists. They simply pick and choose scientific results that support their view only and most of these are being countered by good science. The sceptics do not refer to theses counter arguments and hence are bad scientists…

The book proposal was thus criticised as 'bad science' because lived experience was not seen as a complementary objective knowledge.

The second area where scientists obstruct interdisplinarity is that they feel personally threatened, arguing that they cannot understand social science which they will need to incorporate in their work. For them, standpoint does not matter as science is all about 'objectivity' and in this blinkered view, they are not familiar with the voluminous literature on standpoint arising from race, poverty and gendered discourses.

Yet my experience suggests that the more senior, and more confident the scientist, the more willing they are to enter into a holistic analysis. For instance, an understanding of good and bad science from a distinguished physical scientist (Professor Bhadeshia, Metallurgy, Darwin College, University of Cambridge, UK) is somewhat different. He suggests that 'a good scientist: accounts for prior work, is more interested in the problem than just writing incremental papers, and exposes the work to a wide audience for scrutiny'. A 'bad scientist: rediscovers known phenomena, makes claims before proof and is more interested in publicity than truth'.

He goes on to say that he has certainly worked with interdisciplinary projects which have generated important findings. For example, one team included physicists, mathematicians and chemical engineers. This was on a method called neural networks that the team together adapted for metallurgy, the mathematician in creating a model for hydrogen in steels, and the chemical engineer on surface treatment of steels. He reflects that, 'in real collaborations, we rely on a deep understanding of the 'foreign' subject by the partner. So if there isn't a level of trust and friendship then there may be problems. If you focus on getting at the truth then compromise is difficult, although the same story can of course be expressed in different words by different people'.

Professor Bhadeshia has also worked on a multidisciplinary project organised by Darwin College at Cambridge University where eminent people from social and science disciplines presented lectures related to the theme of structures that are part of our experience of the world. Structures might include bridges, buildings, DNA, the periodic table, flora and fauna, machines, circuits, human beings and their societies, even thoughts and ideas. He later co-edited as a book based on the lectures (Pullan and Bhadeshia 2000). The multidisciplinary interaction mainly happened around selecting speakers and where people needed to be convinced of topics being proposed out of their areas of expertise. This involved long conversations!

In supporting the idea that climate change needs close collaboration between social and physical scientists, he added, 'This is not only good but essential. A lot of science cannot reach conclusions without the context provided by the humanities. I think the humanities have a greater degree of scholarship in what they write and present, in the sense that the quality of writing is nicer, referencing is more thorough and the work is often exposed before publication. The articles are also lengthier. In science, brief stories are published, and perhaps too many papers published, with the style of writing being functional that so many papers are not a pleasure to read even though they may be pregnant with information' (All quotations from personal correspondence, November 2014).

Physical scientists are like anyone else—creatures of their own social contexts and societal structures of race, gender and class. Their mindsets are influenced by the importance society attaches to their subject as well as personal ambitions and insecurities aggravated by social scientists who are pushing into their knowledge territories of subjects such as climate change. The less threatened ones are those who are confident enough to recognise the benefits of crossing boundaries in establishing 'truths.'

This personal story warns ultimately against glib calls for different scientific disciplines to respect each other and work together, a challenge of collaboration that would surely be amplified if lived experiences were added to the mix. 'Joined-up-thinking' about climate change, however, is key to the vision of this book. How it might be brought about requires a thorough investigation of the dynamics of human engagement and the conditions necessary for power relations and diversity to be put to productive use, mutual trust fostered, insecurity and arrogance minimised, and the whole idea of transboundary collaboration institutionalised at a broader scale.

These are the challenges underlying Chaps. 7–10. Meanwhile we discuss the sciences (natural and social) and how they represent climate change further in Chap. 4. This chapter also suggests an interdisciplinary approach that is informed by lived experience to provide a more complete representation. Then in Chap. 5, we turn to another form of knowledge that informs climate change debates—local knowledge. It too has its, at times, evangelical, adherents.

References

Blackmore, R. (2009). Climate change: Causes and consequences. In G. Wilson, P. Furniss, & R. Kimbowa (Eds.), *Environment, development and sustainability: Perspectives and cases from around the world*. Oxford: Oxford University Press.

Breitmeier, H., & Otto, D. (2012). Understanding political processes in climate change negotiations by means of an interdisciplinary curriculum in higher education. *International Journal of Innovation and Sustainable Development, 6*(1), 20–30.

Dyson, F. (1995). The greenhouse effect: An alternative view. In J. Carey (Ed.), *The Faber book of science*. London: Faber and Faber Ltd.

European Commission. (2011). A roadmap for moving to a competitive low carbon economy in 2050. http://eur-lex.europa.eu/resource.html?uri=cellar:5db26ecc-ba4e-4de2-ae08-dba649109d18.0002.03/DOC_1&format=PDF. Accessed February 20, 2015.

Francis, J. A., & Vavrus, S. J. (2012). Evidence linking Arctic amplification to extreme weather in mid-latitudes. *Geophysical Research Letters, 39*(L06801), 2012. doi:10.1029/2012GL051000.

Funtowicz, S., & Ravetz, J. (2013). Post-normal science: Environmental policy under conditions of complexity. NUSAP.net. Available at: http://www.nusap.net/sections.php?op=viewarticle&artid=13. Accessed February 20, 2015.

Giddens, A. (2007). *The politics of climate change*. Cambridge: Cambridge University Press.

Hajer, M. A. (1995). *The politics of environmental discourse*. Oxford: Oxford University Press.

Howard, J. (2011). Environmental nasty surprise, post-normal science, and the troubled role of experts in sustainable democratic environmental decision making. *Futures, 43*, 182–195.

Hulme, M. (2009). *Why we disagree about climate change*. Cambridge: Cambridge University Press.

IPCC. (2013). Summary for policy makers. In T. F. Stocker, D. Qin, G.-K. Plattner, M. Tignor, S. K. Allen, J. Boschung, A. Nauels, Y. Xia, V. Bex, & P. M. Midgley (Eds.), *Climate change 2013: The physical science basis*. Contribution of Working Group I to the Fifth Assessment Report of the Intergovernmental Panel on Climate Change. Cambridge, UK: Cambridge University Press.

Leroi, A. M. (2014). *The lagoon: How Aristotle invented science*. London: Bloomsbury Circus.

Mabey, R. (2013). *Turned out nice again: Living with the weather*. London: Profile Books Ltd.

Midgley, M. (2014). *Are you an illusion?*. Durham: Acumen.

Nelkin, D. (1977). The political impact of technical expertise. In G. Boyle, D. Elliott, & R. Roy (Eds.), *The politics of technology* (pp. 189–205). London: Longman in association with The Open University Press.

Peterson, T. C., Stott, P. A., Herring, S. C., & Hoerling, M. P. (2013). Explaining extreme events of 2012 from a climate perspective. *Special Supplement to the Bulletin of the American Meteorological Society, 94*(9), 564–565.

Petoukhov, V., Rahmstorf, S., Petri, S., & Schellnhuber, H. J. (2013a). Quasiresonant amplification of planetary waves and recent Northern Hemisphere weather extremes. *Proceedings of the National Academy of Sciences of the United States of America, 110*(14), 5336–5341. doi: 10.1073/pnas.1222000110.

Petoukhov, V., Rahmstorf, S., Petri, S., & Schellnhuber, H. J. (2013b). Reply to Screen and Simmonds: From means to mechanisms. *Proceedings of the National Academy of Sciences of the United States of America, 110*(26), E2328. doi:10.1073/pnas.1305595110.

Pullan, W., & Bhadeshia, H. (Eds.). (2000). *Structure in science and art*. Darwin College Lectures. Cambridge: Cambridge University Press.

Screen, J. A., & Simmonds, I. (2013). Caution needed when linking weather extremes to amplified planetary waves. *Proceedings of the National Academy of Sciences of the United States of America, 110*(26), E2327. doi:10.1073/pnas.1304867110.

Stern, N. (2006). *The economics of climate change*. Cambridge: Cambridge University Press.

Stern, N. (2010). *Blueprint for a safer planet: How we can save the world and create prosperity*. London: Vintage Books.

Toulmin, C. (2009). *Climate change in Africa*. London: Zed Books.

UCAR. (2012). Doping the atmosphere. *AtmosNews*, February 6. Available online at: http://www2.ucar.edu/atmosnews/attribution/doping-atmosphere. Accessed February 20, 2015.

UK Meteorological Office. (2014). A global perspective on the recent storms and floods in the UK. http://www.metoffice.gov.uk/research/news/2014/uk-storms-and-floods. Accessed February 20, 2015. Contains public sector information licensed under the Open Government Licence v1.0.

Chapter 4
Representing Climate Change: Science, Social Science, Interdisciplinary Approaches and Lived Experience

Abstract Climate change has until now been represented mainly by the natural, especially physical, sciences. While this has generated impressive insights and placed the challenge firmly on national and international public agendas, it is an incomplete representation (as suggested in Chap. 3), albeit a powerful one among decision-making institutions. Issues arise through the insistence of a deterministic 'scientific method' and with underlying assumptions concerning empiricism and positivist objectivity. Qualitative social science, with its mixed methods approach and positive view of subjectivity to reveal fresh insights, provides a significant complement to natural science with respect to the human dimensions of climate change. The topic, however, is too complex a problem to be represented by natural and social sciences disciplines alone. Drawing on the concept of climate change as a 'wicked problem', the chapter moves towards advocacy of an interdisciplinary approach that draws on disciplinary interfaces and grounds itself in everyday lived experiences. This approach has the potential to be inclusive of diverse voices and their experiential, as well as scientific, knowledge to create a more complete representation of climate change. Issues remain, however, concerning the politics of representation in terms of whose lived experiences are included/excluded and the ways in which they are represented by different actors.

4.1 Introduction

There are several problems with representing climate change, whether this is to one's peers or to a wider public. First, climate change is a hugely complex problem which lacks a clear definition and consensus on either what the problem is or how we can resolve it. Our reaction to climate change is also often based on several internalised, personal questions. We are suspicious of where messages about the problem are emanating from; whose knowledge they are based on; and whose 'truth' they represent when each individual's climate change 'lived experience' may suggest something else. Each individual's 'lived experience' filters these messages and rationalises short and long-term concerns, making it possible for them to be ignored or acted upon.

D. Abbott and G. Wilson, *The Lived Experience of Climate Change*,
DOI 10.1007/978-3-319-17945-2_4

Furthermore, as human beings we are inclined to seek out individual interests rooted in nostalgia, fear, pride and justice that Hulme (2009: 341) talks of. For this, we need answers to questions which go beyond raw scientific explanations. Information we have received 'from above' is often passed down in a linear process that starts with scientific information and moves up the ladder to national and international official bodies who then warn us about the impending doom of future generations and how we must all act. We may or may not be baffled by the science, given a free rein to take or leave action through enforcement, codes of practice or top-heavy policies. However, for us as the public to have a bigger stake, we need to be involved in the representation of the problem and the interventions that are proposed to address it. 'We' here refers to the general citizenry, a citizenry that includes the messages that emanate from the lesser-heard voices of the poor, of marginalised groups, of practitioners—telling our own stories of living with climate change. Capturing the intricacies of this inclusive 'lived experience' allows us to expand our climate change knowledge and provides a scope for a more complete representation of the challenge.

Yet this is easier said than done. The dominant mode of our climate change knowledge is that emanating from natural sciences, as noted in Chap. 3. As such, this can be easier to handle in some senses—for example it is easier to make policy based on quantifiable data. However, arguably, natural science methodologies that make assumptions about 'scientific method' and positivist rational (discussed below) can only tackle climate change in a clinical way and not within the context of local and global societal power relations within which it is set.

Picking up one of the key themes of Chap. 2, ignoring the overall societal context and associated power relations can cost dear and interfere with knowledge dissemination. For instance, our contemporary knowledge of climate science with its roots in the nineteenth century era of industrial revolutions in Europe and the United States remained dormant for a long time. Thus, as described in Chap. 3, the 'Greenhouse effect' was discovered many years ago by natural scientists such as Fourier's (1824) through his work on the heat balance between incoming and outgoing radiation, followed closely by that of Pouillet (1838) and Tyndall (1859). Similarly, the calculations of Svante Arrhenius recognised as early as 1895 that enhanced water evaporation and changes to the surface albedo could double carbon dioxide concentration leading to melting glaciers and global warming. Arrhenius' complex calculations were disseminated to learned societies, but remained on the back burner for four decades, shrouded within issues of proof and validity. Then, in 1938, Guy Stewart Callender's empirical data of the link between a rising carbon dioxide concentration in the atmosphere and global warming went no further than a hostile reception from meteorological professionals. Charles Keeling took up the banner in the 1950s, and the onset of computing power led to modelling of the link from the 1970s. That there might be a problem of human-induced carbon dioxide emissions and the climate system, however, did not enter the mainstream public and political consciousness until the late 1980s.

Clearly then, individuals need an all-round knowledge of climate change that somehow relates to their own experience. This includes both natural science knowledge as well as that of the societal context and power relations that influence the dissemination and action about a given problem. Whilst a social science analysis can also have its own difficulties in representing climate change fully and there are several problems on choosing representative samples when representing lived experience, its strength lies in its ability to untangle individual and societal histories within complex societal power relationships.

This chapter addresses the complex nature of representing climate change. It explores the fundamental principles and underlying assumptions of both natural and social science research methods that are used to generate knowledge that underpins representation, and argues that each has its strengths and limitations. To represent climate change more fully and adequately, it therefore argues for an expanded interdisciplinary approach where capturing everyday lived experiences plays a crucial role.

4.2 Climate Change as a 'Wicked Problem'

The world that we live in has always posed hugely complex questions. Within the twenty-first century these have not diminished and, if anything, appear to have increased—or maybe they were always there but we did not have the knowledge or awareness of them. Thus we struggle to both identify the heart of these problems and find solutions as reflected in our quest to fulfil the Millennium Development Goals (MDGs) that were established in 2000 by the United Nations to address global poverty and related challenges (Chap. 2). For example, one of the most pressing problems the world has today is how to feed the hungry, particularly in economically poor countries where population growth rates are also high.

There are, of course, several other disturbing, interconnected problems we face today. These include, for example, a global concern with drugs, terrorism, inequity, human rights, human displacement, environmental degradation, and of course, climate change—the list is endless.

But if we are to find solutions to these real-life problems, it is important first to identify the nature of the problem. In fact, as Laurence Peter (an educator and management theorist) cited in Conklin (2005: 10) suggests: 'Some problems are so complex that you have to be highly intelligent and well informed just to be undecided about them'. And, indeed, climate change is one such problem where much thought is required to raise questions about it and the knowledge we have to answer these. As soon as we ask one question, this generates further complex enquiries, creating an endless debate on the nature of the climate change problem, let alone the solutions.

Such problems have become known as 'wicked problems'. A small story (sourced from Skaburskis 2008) on how this term was coined relates to the 1960s scenario, when on the one hand the United States was pioneering moon research, whilst on the other its urban social problems were exploding. In the wake of this, West Churchman, a research and systems analyst, was given funding from the US government National Aeronautics and Space Administration (NASA) to explore ways of transferring lessons from space exploration technology systems to solve urban problems. This led to weekly seminars in 1967 at the University of California (UC), Berkeley on technology and social planning where Horst Rittel, a mathematician and the Professor of the Science of Design, argued that there is much difference between scientific problems, which can be 'tame' and those related to social policy and urban planning which are usually 'wicked' as outlined in Tables 4.1 and 4.2 (refined later in 1973). A 'tame' problem can be analysed within existing methods of investigation as characterised in Table 4.1, whilst a 'wicked' problem (social planning, in this case) defies linear approaches and prescriptive methodologies (Table 4.2).

In summary, a 'tame problem' may be technically complicated but have a relatively achievable and stable solution.

As Rittel and Webber (1973: 161) argue, 'The classical systems approach … is based on the assumption that a planning project can be organised into distinct

Table 4.1 The attributes of 'tame problems'

1.	Have a relatively well-defined and stable problem statement.
2.	Have a definite stopping point, i.e. we know when the solution is reached.
3.	Have a solution which can be objectively evaluated as being right or wrong.
4.	Belong to a class of similar problems which can be solved in a similar manner.
5.	Have solutions which can be tried and abandoned.
6.	Come with a limited set of alternative solutions.

Source Conklin (2001: 9–10)

Table 4.2 The attributes of 'wicked problems'

1.	There is no definite formulation of the problem. The problem includes a permanent feedback with its environment. The process of formulating the problem is interconnected with the process of its solution. To ask all the relevant questions to gain information on the problem, you must know all the conceivable alternative solutions.
2.	There are no stopping rules. The logic inherent in the problem does not tell you when to stop the inquiry.
3.	There are no criteria for correctness. There is nothing in the problem to say how the solution should be judged.
4.	There is no immediate test of the quality of the solution.
5.	There is no ultimate test of a solution.
6.	Once committed to a plan of action, change is consequential. You cannot make consequences not happen.
7.	There is no list of permissible operations (If you have a habitual set of actions or prescriptions, try to break out.).
8.	There are no well-defined solutions. You either have many solutions or none. The probability that a wicked problem has one solution is null.
9.	Every wicked problem is unique.
10.	The problem solver has no right to be wrong. Designers are responsible for their work. Unlike scientists, they do not put up work for rejection at a later time.

Source Skaburskis (2008: 278)

phases: 'understand the problems', 'gather information', 'synthesise information and wait for the creative leap', 'work out solutions' and the like. For 'wicked problems', however, this type of scheme does not work. One cannot understand the problem without knowing about its context; one cannot meaningfully search for information without the orientation of a solution concept; one cannot first understand then solve'.

'Wicked problems', on the other hand, are neither true nor false, they have no complete resolutions, they have no moral implications (as maybe the term suggests), but are as complex as the world we live in. There is often little or no consensus to either what the problem is or how to resolve it, and if anything, 'wicked problems' are forever dynamic and active in evolving even newer social dimensions and contexts, coming to life in our attempts to identify the problem itself, rather than solve it!

Within a world full of 'wicked problems', climate change is a 'super wicked problem' according to some (Lazarus 2009: 1153). Thus, in asking whether science can fix climate change, Hulme (2014: 119) argues that wicked problems of climate change defy 'rational and optimal solutions.... And need to be approached differently, obliquely, if its dangers are to be defused'. But how do we do this? Climate change science is still dominated by natural, especially physical, scientists but it is true that there are now considerable inputs to knowledge from the 'humanists' or social scientists, for example in making us aware of the actual and potential human and societal impacts of climate change.

It is increasingly clear that the enormity and complexity of climate change cannot be addressed within singular disciplinary modes of inquiry, especially if we expect people to relate to and act in response of climate change knowledge. We need to take into account: (a) that there are several stakeholders from differing interest backgrounds, each with valid perspectives, (b) that the more the contribution from various disciplines across the natural and social science disciplines, the more the chance of constructing a wider knowledge base dealing with the layers of complexities embedded in 'wicked problems', and (c) lived experiential knowledge may contribute significantly to this wider knowledge base.

4.3 Fundamentals of Natural Science Inquiry and the 'Scientific Method'

'Science' is a broad category which includes 'natural' as well as 'social' science. Natural science embodies those fields of knowledge that deal with objects and processes in the 'biophysical world'—an enormous spectrum that includes, inter alia, the entire solar system within the Milky Way galaxy, the smaller and varied scale atmosphere that surrounds our planet, the geological formation of oil-bearing rocks beneath the sea bed, a colony of mountain gorillas, or even the inside of a wasps nest or termite mounds.

We have traced, in Chap. 3, the fundamentals of scientific inquiry back to the ancient Greek philosopher Aristotle. Here we delve more deeply into natural scientific inquiry. Underpinned by a 'naturalistic approach', such inquiry seeks reliable knowledge through measurement, testing and interpretation. It revolves around the five stages of a 'scientific method' as described by Russell's 1931 (p. 57):

(i) First is 'observation of the significant facts', leading to questions such as, *Why did that do that? How does that work? What effect does that have? What are the processes that makes that happen?* Or indeed, *How can I solve that problem?*
(ii) Second involves 'arriving at a hypothesis' which if positive, will account for the natural phenomenon under observation;
(iii) The third stage is to 'deduce from this hypothesis consequences which can be tested by observation' for prediction;
(iv) The fourth stage includes designing and undertaking observations or experiments to prove or disprove the hypothesis; and
(v) The final stage involves writing up and dissemination of results, publicly disclosed for other scientists to read and replicate.

'Scientific method' is thus characterised by *empirical facts* gauged through direct observation; *systematic,* careful and rigorous planning of test observations or experiments; control checks and *replication* to confirm the reliability of the result; and *verification* of the work to prove validity. Many scientific tools and techniques are used to support observation and experimentation with frequent use of those borrowed from complementary and overlapping 'formal sciences', for example mathematical and computer modelling, and moving towards proof of a hypothesis.

This quest, as embedded in the underlying philosophy of the 'scientific method' is defined in two key elements, (i) the critical role of empirical evidence, and (ii) the significance of 'laws' of physics, chemistry and other science subjects. An important requirement and a basic assumption of 'scientific method' is that all hypotheses and theories must be tested systematically against observations of the natural world and framed within the laws of pure sciences. Hence any claim to scientific knowledge has to be verified through observations, measurements, experiments, data collection. In fact, this is the *only* valid evidence, and anything that is intuitive or instinctive requires 'objective' verification through rigorous empirical testing—a notion that is increasingly challenged by social scientists as discussed later in this chapter.

The fundamental argument behind the scientific method is that science should only be concerned with the natural world and 'facts', and that all hypotheses and theories must be tested against observations of the natural world, measured or observed by an objective researcher in order to justify new 'truths'. This fundamental argument is embedded in the principle of empiricism. It invokes that knowledge is obtained through experience that is acquired from evidence obtained especially through experimentation and which is underpinned by some manner of sensory awareness in its various forms, depending upon the disciplinary strand. For instance, experience and sensory awareness may mean different things for a psychologist compared to a physicist. However, whatever the difference, empiricism considers observation and evidence as fundamental to the establishment of 'truths'. Thus for instance, we moved several centuries ago from the historical 'truth' that the Earth is flat to one that the earth is close to a sphere, a truth that was originally supported by natural 'facts' such as the movement, or otherwise, of the sun/moon, and today is supported by observations from satellite images.

Natural scientists also endeavour to create theoretical explanations based on causal relationships that are determined by universal laws, meaning that, wherever tried and tested, the result is the same. Examples include the theory of gravity, the behaviour of hot and cold air and so on. To defend laws and theories as 'knowledge', scientific method turns to positivism. A positivist evolution is associated with the early 1830s French philosopher Auguste Comte who challenged notions of speculation and utopian ideas of pre-revolutionary France in what he termed a 'negative philosophy'. Instead he argued for a positivist view based on the real world of given circumstances and material objects. Thus positivism focuses on 'positive' facts and actual, rather than imaginary, observable phenomena. In doing so, the approach avoids metaphysical questions (such as those related to being and self) and theological questions (about God) that dominated the era, and which could not be measured scientifically.

A positivist rationale therefore embraces careful measurement and objective data collection to create general laws to predict and explain nature. To obtain objective data, interference from the researcher's personal or political values, which may mar 'truths' and knowledge, needs to be kept in check. Positivist knowledge thus requires propositions or hypotheses and laws and theories to be tested and verified, hence it provides the foundation of 'scientific method'. In fact, pure positivists are so sure about the 'scientific method', they argue that this is the only methodology that can capture the true essence of natural inquiry which may be applied universally. Since Comte's original formulation, positivism has evolved and divided into several strands, two of which are summarised in Box 4.1.

Box 4.1 Logical positivism and critical rationalism

There are several contemporary versions of positivism, with two historically influential ones being (i) logical positivism and (ii) critical rationalism. The roots of logical positivism are in the 1920s Vienna positivists who argued that theories are only meaningful if they could be tested and verified—and this included theories beyond the realm of natural science, for instance, social science theories. Thus they turned to logic to make 'scientific method' more rigorous for verifying theoretical 'truth' or falsity of empirical statements. For them, verifying or falsifying requires propositions to be justified against reality, without resorting to experiential analysis or synthetic statements whose 'truth' needs to be established by empirically tested hypotheses. Logical positivism attracted some important names amongst whom are early logicians such as Ernst Mach and Ludwig Wiggenstein and later Bertrand Russell. Whilst logical positivism contributed significantly to the philosophical debates of logic, language and mathematics in their day, this 'totalising' approach has since been discredited as over-simplistic. We may, however, perceive overtones of logical positivism persisting among many climate scientists, where the validity of models that predict the future under global warming scenarios is established through how well they also describe contemporary and historical climates (see also Chap. 2). As the IPCC 5th Assessment Report for Working Group 1 states: 'The direct approach to model evaluation is to compare model output with observations and analyse the resulting difference' (IPCC 2013: 746).

A challenge to logical positivism came from the critical rationalism of Karl Popper, also a philosopher, in the 1930s. Critical rationalism, as the name implies, is about reflecting on claims to knowledge in a critical and rational manner. To do this, we can only accept a claim that can be proved and/or through reason and experience. Popper argued that empirical evidence cannot be confirmed in advance of the event and thus it is possible that we may come across inconsistencies. Therefore, if the claim is based on empirical findings, it can be tested in order to falsify it. He argued that in testing a theory or other claims, scientists are in fact not looking for evidence, but counter-evidence to see whether or not the empirical evidence is refutable or false. Contemporary critical rationalism elaborates Popper's arguments to several areas of thinking and action around particular problems, fundamentally to replace justification methods with critical rationalism. The United Nations Intergovernmental Panel on Climate Change (IPCC) process, that reviews and assesses the most recent scientific, technical and socio-economic information produced worldwide relevant to the understanding of climate change, may be seen as an exercise in critical rationalism (see Chap. 3, Box 3.4).

Critical underlying assumptions of natural science are thus partially built into the fundamentals of a 'scientific' methodology and partially on a positivist philosophical rationale. The goal of inquiry is to explain and predict. Most positivists would also say that the ultimate goal is to develop the law of general understanding, by discovering necessary and sufficient conditions for any phenomenon (creating a perfect model of it). If the law is known, we can manipulate the conditions to produce the predicted result. In this, the essential premises of scientific inquiry comprise (i) both inductive (where research comes before theory) and deductive reasoning (the theory comes before the research); (ii) value free and objective research; and (iii) validity and reliability.

4.4 Underlying Assumptions of the 'Scientific Method' and Positivist Objectivity

A broad acceptance of the critical underlying assumption that 'scientific method' leads to objective, value free, neutral 'truth' untarnished by personal beliefs, bias or preferences has given natural science an extremely dominant position in society. Powerful institutions within which science has historically resided (such as the Royal Society in the United Kingdom and its equivalents in other affluent countries, and the Christian Church in Europe) have meant that throughout history human populations have been willingly or otherwise obliged to live by its truisms. This has especially been the case where dominant institutions have understood science to be in their service (even if this does mean that claims to be 'objective, value free and neutral' ring hollow). The alignment of science and the Christian Church, for example, ruled lives in the Europe of the Middle Ages (475–1550 AD) when the Church controlled all scientific and philosophical ideas about the natural world. Through this control, it established 'undeniable truths' about biology, physics and astronomy based on the Biblical interpretations and the teleological reasoning—meaning that everything is to be explained by the ultimate purpose it serves (Chap. 3, Box 3.2)—which was so distinctive of the philosophical system of the ancient Greek philosopher, Aristotle. This was a powerful doctrine. It was not until the 'scientific revolution' (circa 1550–1700) that new ideas in physics, astronomy and human anatomy began to challenge and undermine the laid-down laws and views of the church and the Aristotelian system of teleological reasoning. Two publications in particular were seen as the tipping point for the start of the revolution: Nicolaus Copernicus's *De revolutionibus orbium coelestium* (*On the Revolutions of the Heavenly Spheres*) which shifted from the geocentric model and placed the sun at the centre of the planetary system and Andreas Vesalius's *De humani corporis fabrica* (*On the Fabric of the Human body*) which challenged existing Galen's theory of humours through observation and meticulous drawing of the human body, its skeleton, muscles and bones. Science carries this legacy of establishing 'truth' based on an 'objective' method of inquiry even today.

Notwithstanding recently documented evidence of falling public confidence in science (see Sect. 4.6 below and Chap. 7), it has also historically been, and largely continues to be, dominant in that it is perceived to work, that is, it provides solutions to problems—it delivers the goods. This might be through direct application, such as in engineering, but also through providing evidence for policy, as does climate science. The issue here then does not ultimately concern science, but the definition and representation of the societal problems that it is marshalled to solve?

To give one pertinent example, associated with the continuing growth of a capitalist, industrial society mainly through Europe and the United States in the late nineteenth and early twentieth century was a continuing problem of finding a more effective mode of assisted transport than a horse and cart, and which was more flexible than the steam railways. The internal combustion engine, based on the scientific laws of thermodynamics, was invented in 1868 by the German, Karl Benz and led to the automobiles that we know today. Secondary impacts were conceived only through comparison of the automobile with the horse which it sought to replace, the former appearing much cleaner and easier to maintain. With the advent of the automobile, the smell and general health hazard of horse dung throughout the streets of London and other big cities would become history. Nobody at the time asked about the possible health hazards arising from pollution of the atmosphere through automobile emissions of lead, noxious gases and particulate matter, or if they did, which might have been the case in sci-fi novels, nobody took them seriously. Moreover, nobody wanted to take them seriously as, with increasing scientific knowledge, the future could look after itself. Certainly nobody asked whether increasing carbon dioxide in the atmosphere from internal combustion engine emissions might contribute to potentially harmful global warming.

The issue, therefore, does not always concern the power of science in public institutions per se, but the problems it is asked to solve. These are not necessarily completely wrong problems, but incomplete problems in that they have been defined in function of the imperatives of dominant societal concerns of their time. The ultimate power lies with those who are able to represent those dominant societal concerns—at the turn of the nineteenth/twentieth century the capitalist leaders of industry supported by the political class.

Whether science is seen to work or not, there are, however, several ways in which both the 'scientific method' and the 'truths' thus established may be challenged. To begin with, take the notion of objectivity—ultimately devoid of moral and political values—as central to empiricist and positivist approaches in natural science. To generate a research process that is rigorously designed and controlled, the researcher is expected to observe, record and analyse empirical findings neutrally at every stage in order to discover universal laws and 'truths'. This does, however, beg the question of whether knowledge can ever be obtained in such a detached, objective and neutral manner. Every scientist is human, and no doubt has personal opinions, things in which they believe or disbelieve as individuals, their own favourite topics and interests which influence how they approach and see the world, all of which influence the questions they ask. This is clear, for instance with ethical dilemmas when doctors with strong personal moral beliefs will not research on human embryos. So, whatever the underlying philosophy and a legacy of how they or their mentors should carry out 'objective' research based on 'objective' criteria, it seems likely that internalised beliefs and a 'human factor' will still have a leading influence.

As societal creatures, we inevitably operate in a cultural environment that holds certain beliefs and doctrines, and researchers are not exempted from that! For instance as mentioned above, historically religious, absolute 'truths' dominated scientific thinking and processes in the Middle Ages. More recently, biologists have measured intelligence and head sizes in establishing 'truth' and evidence of racial inferiority and supremacy within their own cultural environment. Fortunately this theme has now faded (although not completely disappeared) from popularity. Further examples include the work undertaken by scientists working for the tobacco companies in the United States who have sought to discredit the scientific link between smoking and cancer, and those scientists who have questioned the link between anthropogenic carbon dioxide emissions and climate change. Arguably, because large multinational corporations have often funded much of this research, it has focused on finding a negative link rather than a positive one. Therefore, however hard scientists seek objectivity, many factors including who is paying the piper and the cultural context of the time within which human values are defined, impact on how hypotheses, research design and questions are represented and analysed.

A second but equally important assumption is based on the issue of validity and reliability. All good science aims to be valid and reliable. In natural science, valid research is that which accepts an empirical or positivist approach as discussed above, accepts the entire experimental concept and establishes whether the results meet all of the requirements of the 'scientific method'. We should note that there are two aspects of validity to consider, internal and external validity.

Internal validity dictates how an experiment, observation or survey is designed. Internal rigour is an absolute necessity if valid results are to be produced. This requires an ordered and careful approach to data collection. External validity is further down the line and is concerned about how one processes the results and asks questions of them to discover causal relationships. These relationships are easier to validate if there is some form of control built into the experiment, for example the use of statistics including randomised samples or locations to determine significance and difference. Note the term 'significant' is not necessarily evidence of an 'absolute truth'. Also correlation does not prove a relationship because there may always be some other reason, not yet measured or observed, which has contributed to the results.

Reliability refers to repeatability, comparison and consistency of a result. It is reinforced when other researchers are able to conduct exactly the same experiment, observations or surveys, under the same controls and obtain the same results. The use of instruments for measuring, for example, a stop watch, is also seen as a robust way to collect accurate and reliable data. In natural science, following a deductive approach, we can say that the more repeatable and consistent the finding, the more weight it adds to a hypothesis which eventually perhaps becomes recognised as a scientific law or 'truth'. As noted in Box 4.1 above, climate change models, for example, are tested against past climate data to check for validity and reliability and many modifications are on offer on how to improve them. See Fig. 4.1 where the IPCC (2013: Chap. 9, Fig. 9.2) checks the accuracy of an ensemble of climate models against historical observed temperatures. Note for our purposes here that the relevant Figure is (b) where the shading represents the difference between temperatures derived from the ensemble of models and those observed. Mostly this is 1 or 2°, with exceptions especially around the polar regions where good data are less readily available. Yet, of course, none of this accounts for social and cultural contexts where reliability is gauged in other ways—through, for example, our lived experiences.

There is little doubt that whilst science continues to strive towards establishing 'truths', every now and then we are shaken out of our present understanding of the natural worlds in large jumps. Thomas Kuhn, a physicist who wrote many works on the history of science from the 1950/1960s onwards, characterised these jumps as paradigm shifts or revolutionary science. This is science which somehow challenges its previous basic theories and assumptions. Science history is full of examples of how and when this has occurred, including Charles Darwin's theory of natural selection and Wegener's Plate tectonics theory for large-scale geological events. Paradigm shifts are a result of challenge to the established view of reality and 'truths', and arguably, science today continues that challenge, for example with the discovery of the DNA structures in the 1950s which has revolutionised our thinking and enriched forensic science knowledge in leaps and bounds. Indeed, these paradigm shifts are the large-scale embodiments of what many today understand to be the process of science—which is that at all scales it proceeds through challenge (see Chap. 3 and Hulme 2009: 75).

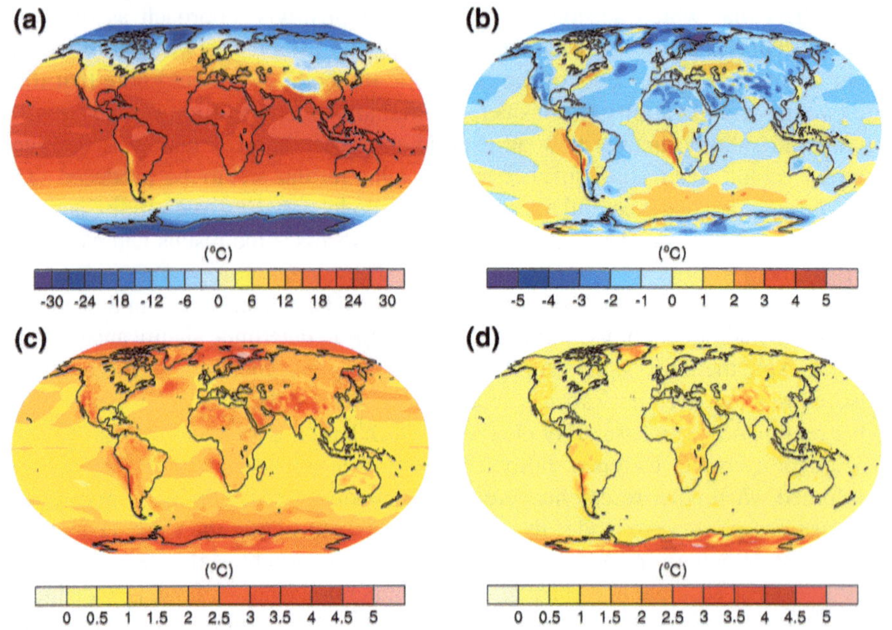

Fig. 4.1 Annual mean surface (2 m) air temperature (°C) for the period 1980–2005. **a** Multi-model (ensemble) mean constructed with one realisation of all available models used in the CMIP5 historical experiment (CMIP5 is phase 5 of a model comparison project). **b** Multi-model-mean bias as the difference between the CMIP5 multi-model mean and the climatology from ECMWF reanalysis of the global atmosphere and surface conditions (ERA)-Interim (Dee et al. 2011); see Table 9.3. **c** Mean absolute model error with respect to the climatology from ERA-Interim. **d** Mean inconsistency between ERA-Interim, ERA 40-year reanalysis (ERA40) and Japanese 25-year ReAnalysis (JRA-25) products as the mean of the absolute pairwise differences between those fields for their common period (1979–2001)

Nevertheless, with its history of establishing 'truths', natural science continues to be powerful and almost expects to be backed by powerful institutions and lawgivers, making it difficult for the general public to challenge its findings. That said, the public is not a passive recipient and citizens, drawing on their own lived experiences, will continue to challenge science and its underlying assumptions on which many findings are based.

4.5 Fundamentals of Social Science Inquiry and Power Relations

'Social science', as the name implies, is the study of the 'social' or society and its main focus is on societal power relationships arising from all kinds of specific micro- and macro contexts, and the connections between them. It is therefore a very broad term encompassing several disciplines, including sociology, anthropology, geography, history, political sciences, economics (to name some). A social science analysis stretches beyond theory and is applied to societal problem-solving through policies, for instance in social work, psychology, language analysis and linguistics, educational action research and international development.

The 'science' aspect of social sciences refers to academic study which is carried out systematically and analytically to produce some generalisation with similar rigour and testing of validity as is applied to natural sciences discussed above. Because of the complexity of society and that no human being or event is exactly the same, social science cannot make judgments based on any singular research method. Most social scientists will therefore borrow a range of methods from varying disciplines, tending to draw on a multi-method approach for both research and 'finding out' as well as the analysis of those findings (as some natural scientists also do, but more so). For instance, 'finding out' may involve observation and participatory techniques based on anthropological methods and surveys, statistical analysis based in mathematics, structured and semi-structured interviewing based on sociology, digging deep into historical records from archaeology and history, along with analysis of drawing, sculpture, arts and music from the arts and media schools. What is ultimately chosen will of course depend on the nature of the enquiry. It is this reliance on mixed modes and 'untidy' data that historically social science has been attacked for not being rigorous enough as 'true' science is.

For social science, society revolves around the relationship an individual has with others at both micro- and macro levels within their immediate and wider surroundings. It very much concerns power relationships between people and groups of people and what defines these, no matter whether they are apparent to the observer or are unseen or unmentioned but ever-present (like the proverbial elephant in the room), shaping both the individual's and societal behaviour. Three 'giants' who have highly influenced and shaped social science analysis of power relations in society are Emile Durkheim (France, 1858–917), Karl Marx (Germany, 1818–1883) and Maximillian (Max) Weber (Germany, 1864–1920).

In brief, Durkheim in his pioneering study *Le Suicide* (1897) established several correlations between 'social facts' such as religious values, gender (men and women's socially constructed roles) and peace and conflict to study suicidal behaviour in a particular society at a given time. By doing so, he took a very private individual affair (that of suicide) and gave it a very public, societal context to draw generalisations. The analysis of 'social facts' is therefore critical to an understanding of how the structures and hidden 'laws' that govern a society work and what it expects.

Marx was also concerned about correlations within society but, in the epoch of the early industrial revolution, he added the critical dimension of class to his analysis of society. For Marx, the essence of society lay in a 'class struggle', located within the powerful 'bourgeoisie' who own the means of production and the 'proletariat' who sell their labour. He traced class struggle to antiquity arguing that 'the history of all hitherto existing society is the history of class struggle' (the opening line of Marx's Communist manifesto 1848). For Marx, therefore, history, politics, economics and social behaviour are very much intertwined and manifest in societal power relationships as defined by individual class position. Whilst Marxist ideas have historically influenced revolutions, uprising and worker solidarity all over the world and continue to do so today, for social scientists, Marx has offered a colossal, inescapable dimension to the analysis of social values—that based on power, class and underlying ideology. And, even today, argue as they might, social scientists cannot escape a class analysis in any societal enquiry!

Within the context of a more modern world and the rise of global capitalism Weber, the third name associated with social science, offers further dimensions to societal analysis, including religious and political party affiliations. In his seminal work, *The Protestant Ethic and the Spirit of Capitalism* which is a series of essays compiled between 1904 and 1905, Weber argues that capitalism has grown since the onset of the industrial revolutions in Europe and North America because of the filtering of religious messages of a 'protestant ethic' which celebrates hard work as a path to riches and spiritual salvation. He adds that, the ramifications of industrialisation and capitalism may lead to violence in the form of war and conflict, and also a growing system of classification of populations by state bureaucracies, a system that includes several sub-categories of citizen classes stratified socially by their wealth and status in society.

To this day, we continue to analyse exhaustively the works of these three 'giants' and their interpretation to societal power relationships within modern contexts. Whist acknowledging the contribution made by natural scientists to climate change knowledge, social scientists see this as technical, but lacking in a power analysis.

4.5.1 A Shift from Numbers to Qualitative Data

Social science is thus a systematic study of society that is located in several disciplines and has at its core an analysis based on variables derived from the historical, political and economic make-up of society. As noted above society and power relations are so complex, social science cannot make judgments based on any singular research method alone. Many social scientists, therefore, adopt a multi-method approach. What method you choose will, of course, depend on the nature of the enquiry.

What marks out many social science assumptions are (i) a methodological stance which, whilst also using quantitative data, to a significant extent focuses on qualitative data as a primary tool for commenting on societal interaction, and (ii) an ontological one which not only regards research as value laden and anything but objective (in contrast to positivism of natural sciences), but one that actually celebrates subjectivity of research as of value in itself. There are, of course, exceptions. For example, typically economics is based on quantitative analysis and also claims to be objective.

Social science research has historically not always been influenced heavily by qualitative methods. In fact social science continues to draw on quantitative data and ponder how it can be worked within a convincing interpretation that is essentially located in a positivist, statistical framework (see for instance, Bulmer and Warwick 1993). Since the later part of the twentieth century, however, qualitative data been increasingly used in the public domain to influence decision-makers and public opinion.

A marked input into 'what type of data' has evolved from gender studies, essentially challenging the supposed gendered leanings and ideology of rationalist,[1] quantitative methods. Following the 'waves' of western feminist discourses in the 1960s and 1970s, publications on the gendered nature of research processes began to appear in the early 1980s. What was new and radical here was the notion that research is not simply about gathering data and analysing it according to set models and formulae and getting the techniques right. Research, it was argued, is much more about the hidden meanings and assumptions behind all stages of the research process and reflects our societal beliefs, gendered relationships, and questions of equality, inequality and discrimination which are entrenched within the dominant political, economic and ideological contexts of the world we live in (see for instance, Roberts 1981).

The notion that data is ideologically loaded is now accepted within many other disciplines within the broad field of social sciences. Whilst the unyielding conflict between which type of data is better for discovering 'truths' has somewhat died down, there are still some dismissive attitudes from time to time. For instance, Berg (2004: 3) cites a Miles and Huberman quotation (1994: 90) from Fred Kerlinger, a quantitative researcher, that 'There is no such thing as qualitative data. Everything is either a 1 or 0'. By and large, however, there is an acceptance that there is value in seeing both data types as complementary rather than opposites.

4.5.2 Underlying Ontological Assumptions and Celebration of Subjectivity

As discussed above, whilst all processes of natural science research methodology are based on the notion of objectivity and neutrality, contemporary social science mostly argues that nothing is free of a social context and existing social knowledge of the time. Attempts to mask true motivations and feelings with respect to research choices are considered a form of 'self-deception', and that an emphasis on rationality, neutrality and attempts to objectify can be disingenuous. In fact, researcher subjectivity is a necessary part of social research and has to be recognised explicitly and reflected upon carefully to produce fresh knowledge and give

[1]Rationalism has a long history, being associated with the seventeenth Century Age of Enlightenment, associated with philosophers such as Descartes, Spinoza, Voltaire. Simply put, rationalists draw on reason and its objectivity rather than emotions, experiences, subjectivity to explain and enhance knowledge.

a fuller account of the study in question. Rather than spending considerable time and effort in weeding out subjective 'interference' in a search for pure objectivity, we should in fact celebrate subjectivity. Thus since the 1970s there has been a growing literature on the importance of recognising personal feelings and motivation behind research, and an encouragement on subjective reflection throughout the research process.

Sometimes personal feelings and subjectivity—themselves born from lived experiences—tell you more about a society and what is happening therein than technical, statistical data. To emphasise this point, Box 4.2 is in an extract from an earlier article one of the authors of this book wrote (Abbott 1995: 9–10) while working in the slum localities of Mumbai when researching poverty as a real-life issue, in other words, moving away from economic data, such as Gross National Income or Gross Domestic Product per capita.

Box 4.2 A subjective reflection on poverty

The question of being a foreign researcher has, in fact, posed a lesser dilemma for me than my ability to comment on poverty. Amongst other things, I continue to worry about my (comparative) privilege: How could I, who would possibly never experience poverty at this level, write about it? I also continued to be worried about exploitation: Have poor people not been 'research objects' for long enough? Yet I cannot answer their question, 'What will you do for me?' when it was me who was 'after something' (my research perhaps?).

Nothing in social science had prepared me for dilemmas like this, or taught me how to research poverty. My privilege kept on interfering and in their friendliness and hospitality, the women (in the slums) continued to remind me. Thus, they rushed to borrow a chair from a neighbour when I visited, even when I insisted on sitting on the floor; they sent children out to buy soft drinks and milk even when I knew that these would never be bought for the children themselves; and they insisted on 'honouring' me by cooking a 'mutton curry' even when I am a vegetarian because this is a rare treat for them!

Through their actions the women made it quite clear that there was a vast gap between us and that I would never be able to understand poverty except from a privileged position. Therefore… my understanding of poverty came through my own reactions—an understanding which shocked me out of any assumptions I held about being poor and which, in turn, generated numerous other moral dilemmas.

To give an example, social science had taught me that when conducting fieldwork, expect the respondents to ask questions about yourself and be prepared to give truthful replies. So when women asked me how many children I had, I replied that I had two. What came to me as a shock was their next question, 'How many have died?' The matter of fact tones of this question (which occurred several times) and my own startled response to something I would never have dreamt of asking anyone at home (the UK), told me more about poverty (and my privilege) than any amount of quantitative data on the subject of infant mortality. In another instance, on a repeat visit to one woman, I saw her seven-year-old daughter lying on a mat. I stroked the child's head and asked her what the matter was. Her mother informed me she had typhoid. I had to prevent myself from jumping and showing my panic at the thought that I had actually touched someone with typhoid, thinking I too would catch it. Having tried to control myself, I asked her if the child had received any treatment. The mother nodded and showed me a packet of the painkiller, paracetamol.

In my supposedly objective role as a researcher, I faced many dilemmas in a matter of seconds. I cannot comment on how the mother must have felt because I had not faced that type of hopelessness. I could not inform the mother that the medicine on which she had spent (what was for her) a lot of money, would do little for the child. I could not 'rescue' the girl because in poverty these things happen over and over again. Yet I could not but read into the mother's eyes that she wanted me to give her some money for the child, and I could not but empathise with her because I had a child of about the same age. There are no sociological guidelines on how to deal with this or the 'guilty' feeling I developed because I could walk away at any time, and the people who had been so friendly to me could not.

Whilst there is plenty of quantitative data available on poverty in big city slums, this kind of qualitative data allows one to move away from the need for objectification because here, subjectivity and subjective reflection which draw on one's inner feelings help to throw light on multifaceted, complex concepts of privilege and poverty. They also help one understand the interaction of phenomena such as climate change with these complex concepts. For example, Box 4.2 refers to a child suffering from typhoid. Box 4.2 does not tell us the exact source of contracting the disease in this particular instance but we know in general terms that it is a consequence of drinking contaminated water. To extend, now hypothetically, to possible links with climate change, the contaminated water may have been drunk because the commodity is expensive and in short supply, a problem which would be exacerbated by a drought. Another example (again hypothetical to this particular instance) is that children who have died may have fallen into a swollen river after a flood.

In contrast to rationalist arguments, some social scientists argue that if you use methods that seek distance, your understanding of the topic will also be at a distance (Patton 2002). It is for this reason, and the fact that some topics in themselves are highly poignant societal issues, that it is difficult to ask the researcher to be distant. In fact a critical discourse on the notion of subjectivity and subjective reflection has become embedded strongly within certain social science disciplines—women's studies, cultural studies and psychology, for example. This is perhaps because the topics that these disciplines tackle are very contentious in themselves. Examples include domestic violence, sexual deviance, forced marriages, mental health, suicide, self-harm, racism and so on—usually topics that are not openly addressed by society, or indeed by the individual.

The subjective in research is also important because it enhances the researcher's own theoretical understanding of the personal. For instance, people often do not know how to deal with disability, even if the disabled person is fairly confident in dealing with their disability. Here is a reflection from Bridget, who is herself disabled, with a subjective reflection on her PhD fieldwork experience in Uganda:

> Research embodies the prejudices of society towards people with disabilities, but also the web in which the research sits is connected by normative hegemony (Box 4.3) and thus creates the abnormal, the 'other', the disabled, the native, the colonised subject. Thereby the research replicates hegemonic notions of normativity. Could I, as a disabled researcher, break the 'normative hegemony' which disrupts people's prejudices and asymmetrical structures of the norm? The elite don't like it, they try to push you down, or they treat you as non-threatening, or trivialisation. Differing disabilities are regarded more negatively than others and this connects to complicated social forces. For instance, disability is connected to the ability to produce in industrialised society, where some impairments are more disabling than others.

Box 4.3 Normative hegemony

We live in a world where we define ourselves by what is 'normal', and therefore acceptable within our society. For example, we are ranked (and even classified sometimes) as having lower or higher than normal intelligence, mental capacity, height, weight—all of which are constantly measured and assessed. These norms are often set by those with power (such as the fashion industry, diet industry) and reflect societal values.

A disabled person is often compared with those who have a 'normal body' (whatever that is!). This is sometimes done wittingly or unwittingly as we tend to internalise societal everyday norms. The disabled person is thus considered 'a problem' in a societal construction of normalcy. The issue is: what is the problem? Is it the disability or the dominant ideology that constructs normalcy which sidelines people with disabilities?

Similar arguments over normative hegemony can be made over other societal issues such as sexual orientation, race and gender.

We, the authors of this book, can add a climate change dimension to this quotation from Bridget. Surviving extreme weather events often means being able to physically move to a safe place immediately. For example, dramatic television images of floods in economically rich countries show a distinct bias towards having to rescue old people and those with physical disabilities. The situation is exacerbated many times in poor countries where the infrastructure for rescue might not be available. Disability here is connected to the ability to move very quickly, giving a whole new meaning to lived experiences of people who are not able to do so!

4.6 Interdisciplinarity, Representation and Climate Change

The question then is not only about which methods we use to capture and represent what is meant by climate change, its impacts and responses, but how the underlying assumptions of disciplines shape the information we are fed. Social science, as demonstrated above, seeks to bring issues of power and privilege into the forefront of climate change debates. Pertinent questions revolve around those in poverty and marginalised groups who will be lesser able to deal with both short and long-term impacts because their personal entitlement and capacity is low (see Chap. 6 and our citation of Amartya Sen's capability analysis), or their Governments are too resource-poor, lack capacity or the political will to act. Who gets a say is a political act.

There are several issues here. One concerns the continuing privilege of natural sciences to inform us when public bafflement and scepticism of climate science suggest it is an incomplete representation of the overall problem. People worry about the trustworthiness of scientific research and facts, especially with media interest in the 'climategate' scandals such as that at the University of East Anglia (UEA, UK 2009). This involved highly reputable scientists who were accused of having manipulated data to support the popular thesis that anthropogenic activity is largely responsible for climate change (Hickman and Randerson 2009). The damage of these allegations to both the University and scientific community along with their findings was so high that the UEA commissioned a special independent inquiry and report (Boulton 2010). The scandal involved 1000 emails from the UEA Climate Research Unit (CRU) which were hacked into and made public through mass-circulation media. These threw doubt on the work of the CRU, and generated scepticism about the honesty of climate science generally, including that of the United Nations Intergovernmental Panel on Climate Change (IPCC) just prior to the important Copenhagen 'Summit' of world leaders on climate change. There were various concerns, ranging from the conduct, honesty, rigour, openness and unethical behaviour with which the CRU managed its findings and how these were presented to the public (Ibid: 10). In this particular case the CRU scientists were in fact exonerated on all counts and found innocent of unethical behaviour. However, arguably the media was keener to report the scandal than the ultimate findings leaving an imprint of doubt on the public mind.

Second, is a concern about ethical and moral issues surrounding the contemporary climate change debate (see also 2.5 on lived experience in relation to culture and value). Ethical issues are, of course, closely linked to morality and intrinsically located within a philosophical framework of values. Ethical considerations attempt to guide right and wrong societal behaviour, offering a systematic, acceptable code. Take for example the topic of climate change. What are the moral and ethical dilemmas that this poses? We as human beings are aware that something is happening in terms of climate change—whether or not scientists agree or disagree as to the cause. We are therefore aware of the millions who are displaced and suffering every day as a result of drought, flooding and other extraordinary climatic events. Information at our fingertips, for example the Famine Early Warning Systems (FEWS) gives us a clear indication of how many people are on the brink of chronic hunger and starvation because (alongside war and conflict) of climate events. At the same time we are aware of debates about geoengineering of climate change, carried out to manipulate and control regional variations and find solutions, for example through rain seed cloud intervention (a technology-led solution which may, of course, create further problems). There is also strong credibility in the argument that the richer, consumer oriented countries contribute much more than poorer countries to wastage, release of carbon gases and the greenhouse effect. Even though we recognise the global nature of climatic linkages, as a world we are divided in our answers to the questions, 'Whose responsibility is it?' and 'What should be done about it?' Meanwhile, at a personal, individual level, we make moral choices either to curtail our consumerism or seek out answers to why we should care (see also Chap. 6 and the case study of students comparing their individual carbon footprints). Issues related to climate change and lived experiences of food security, consumption and individual and social behaviour therefore generate many moral, behavioural and ethical questions.

Third (but not finally), arguably, as an academic and practitioner community we are not always sufficiently equipped to deal with 'wicked problems' because, at the least, we tend to:

(i) Compartmentalise knowledge, often confining this to associated sectors (such as industry, energy, water, agriculture) or to particular disciplinary ways of thinking (physics, economics, politics, sociology).
ii) Create artificial subject/disciplinary boundaries (for example between human and physical geography) even though natural and human interactions need a holistic analysis.
(iii) Lack sufficient collaborative links between researchers, practitioners, professionals and general civil society. We do this for many reasons which are not always immediately apparent to us. We need to make space to think about how we can address issues of knowledge construction based on rational, inclusive approaches instead of the narrow knowledge construct offered through 'expert' and 'professional' groups which are often confined by their segmented institutional specialisation, rules and regulations, bureaucracies, resource ceilings, and generally their individual motives and interests.

Climate change is essentially located in questions of environmental sustainability and the problems that we have in part created through high industrialisation, uncompromising technology a rush for mass consumerism and exploitative behaviour. It consists of highly dynamic systems in which natural scientists play a critical role in providing proximate explanations. To complement, those in the fields of social sciences seek human contextual explanations for the global dependence on energy as the world embraces a neoliberal capitalist ideology, and the challenges faced by policy makers for action on mitigation and adaptation. Within this is a crucial recognition that in an interconnected and a linked world, rich country consumerism and energy consumerism will impact lives and livelihoods also, if not particularly, in poor countries. Climate change impacts will be spread across all humans everywhere, but will be felt unevenly, influenced by the social relations of wealth, power and inequality.

Thus, to represent a broader and deeper climate change knowledge, there is a strong reason to take an imaginative leap into interdisciplinarity and create a collective knowledge base that reaches out beyond defined disciplines in a unified manner. Interdisciplinarity is not anti-disciplinary. Rather, it embraces disciplines in new ways because it is grounded in where people are at in relation to the societal problems that surround and, for many, threaten to engulf them, not some abstract idea. Interdisciplinarity relates to the interfaces between disciplines as the compound word infers. It accepts and acknowledges the depth of disciplinary understandings, but contends that significant gains in knowledge and understanding come from their interaction at the interfaces. Interdisciplinarity thus brings together disciplines in a common endeavour to generate new knowledge out of their disciplinary overlaps, drawing on the concepts of the contributing disciplines but is not bound by them.

In other words, the super 'wicked problem' of climate change, cannot be dominated within singular disciplinary fields, but requires a more holistic picture drawing on both natural and anthropogenic related disciplines. It allows both science and lived experience to work together to recognise gaps in knowledge and how those who often lack voices know more than those in power as the example in Box 4.4 illustrates.

Box 4.4 How waste pickers of mitigate climate change for Dehli

Chintan (an Indian NGO working with waste pickers) estimated in its report 'Cooling Agents' (2009) that Delhi produces 9000–10,000 metric tonnes of waste per day, with a solid waste forecast of 23,000 metric tonnes per day by 2020.

Like in many a megacity, there is an army of waste pickers, usually rural landless migrants who are waiting for the waste rich people make. These men, women and children from low Hindu castes or Muslims, living on the extreme edges of poverty, are the outcasts and 'untouchable' segregated members of society that Delhi wants nothing to do with.

There are, of course, detailed practice guidelines in the policies on Waste Management at a national level, e.g. the Ministry of Environment and Forests' Municipal Waste Management and Handling Rules, 2000. At a local level, these are often shaped by inefficiencies and malpractice. They lack a holistic approach and direction on how best to manage waste in the city's interest through its many layers of bureaucracy.

The waste pickers of Delhi thus drive the city's recycling efforts, keep the streets clean, save the municipal corporation huge sums of money and, in turn mitigate large quantities of Green House Gas.

Yet they continue to face police and civic harassment being considered a nuisance, scavengers and eye sores. In fact, they know very well the informal sector hierarchies and their contribution to recycling of solid waste is phenomenal. For example, The 2009 study estimated that informal sector waste pickers collect 15–20 % of total Delhi's waste by weight and recycle glass, metals, plastics and paper alone reducing emissions by an estimated 962,133 tonnes of CO_2 each year (roughly equivalent to removing 176,215 passenger vehicles from the road annually or providing electricity to about 133,444 homes).

To enable policy shift for environmental justice, organisations such as Chintan therefore undertake rigorous research requiring extensive consultation and discussion together with multi-level dissemination of knowledge (http://www.chintan-india.org/index.htm accessed 05/01/2014).

It is, thus, to interdisciplinary inquiry on climate change that lived experiences may make a significant contribution. Whatever other limits that exist with respect to lived experience (see Chap. 7), disciplinary boundaries are not one of them. It is in this sense encompassing. It expands further the knowledge base and provides voice to those who are usually under-represented. Because it is by definition grounded knowledge in everyday problems, it provides a focal point for generating fresh insights from the interfaces of disciplinary boundaries.

4.6.1 The Politics of Representation

Nevertheless, there is a problem with how one represents climate change and the lived experience of it. The politics of representation concerns whose voice is heard by those who have power as decision-makers on interventions to meet the challenge. Thus, professionals, scientists, others with specialist knowledge from academia and industry, and many more all vie to be heard and influence policy and practice on climate change. It becomes a Pandora's Box, opening up to discontent and conflict between various interest groups. These groups comprise orthodox political institutions such as parliaments that legislate and generally represent citizenry, public and private sectors, and a huge and varied network of NGOs and activists, all of them threatening and pushing their own interest.

Moreover, these issues of representation arise within climate change stakeholder groups as well as between them. For example, criticisms have been levelled at the IPCC over its choice of experts to review the climate change literature on which its periodic assessments are based. In response, for the latest 2013/2014 Assessment Report the IPCC increased its number of coordinating authors from 559 (for the previous 2007 report) to 831. Over 60 % of this number was new to the IPCC process and 30 % came from less affluent countries. Whether not these changes are sufficient to ward off further criticism, however, is a matter of conjecture.

Representation issues are therefore ubiquitous in climate change debates and decision-making processes. In these circumstances, complex and broad issues such as those associated with climate change become clouded and often their discussion is no more than a game. Populist and opportunist games are played between politicians, the media and personalities.

Below, for example is (an everyday story?) of a water deficit emergency and representation of this. In his book *Everybody Loves a Good Drought* based on a series of reports for the *Times of India* by freelance journalist Sainath (1996), he suggests how droughts are treated as events where everyone moves on once the event begins to attract lesser media interest instead of the complex processes that affect long-term livelihoods and experiences. He argues that, whilst drought is amongst the most serious problems that India faces, the poor are left out of the processes of consultation and participation with the result that 'drought relief is rural India's biggest growth industry…and often, there is little relation between the two' (Ibid: 317). He goes on to show how emergency relief institutions including Government and State agencies, and foreign donors do not always work in the best interests of the water-needy. This is often inadvertently done as they are quick to give large funding contracts to profiteering (not always efficient) private firms with alternative interests to build infrastructure, maintain water tankers and dig wells. Powerful local hierarchies also act quickly to colonise water resources and set up lucrative trade, selling and marketing water to those who can afford it. The relief thus rarely touches the lives of the rural poor directly affected by water scarcity.

The lived experiences of the poor, mainly tribal and lower caste groups who have to manage and adapt to prolonged droughts, are miserable, bitter and desperate. Any thoughts they have of the future are of impending famine. In Kalahandi (Orissa) during 1985–86 there were stories of mass migration and hunger recounted as follows: 'Here is a picture of hell…all those who were unable to get away were dead or dying …For food they pick up poisonous roots and leaves—the only things that grow there'. Moreover, while the rainfall data of the area suggest otherwise, the villagers spoke of the major 1965 drought as going on for 20 years (ibid: 337).

Things grew critical and stories emerged of women being sold to the brothels of the neighbouring city of Raipur. For example, in July 1985 in the Nuapada (previously in the Kalahandi) district of Orissa, a 14-year-old tribal girl, Bonita, was sold by her sister-in-law, Phanas Punji, for 40 rupees (about two-thirds of a dollar or just over half a Euro) so that she could feed her two starving children. This made national headlines. Shocked, the then Prime Minister Rajiv Gandhi visited the district and announced a development package called the KBK (Kalahandi-Bolangir-Koraput) Development Scheme for the district. However, the scheme was soon forgotten when funds ran out (Gupta 2003).

This example clearly illustrates that both media and populist political interest in events wane very quickly but the poor continue to remain invisible within the many consequences of events. This is not to say people are passive victims. The politics of representation is evident as they demand to be heard in the anger they express, their activism and through the literature on climate justice. This literature includes a huge cache from eco-feminists all over the world but particularly from the poorer countries. The most prominent of these is the well renowned Asian activist and author Vandana Shiva who has fiercely criticised and led the fight against globalisation, the development of biotechnology to serve mainly corporate interests rather than attack disease and hunger, the World Trade Organisation and for environmental and social justice, especially for women and the world's poorest. In Africa, leading the Green Belt movement in Kenya and women's struggle, Wangari Maathai was awarded the Nobel Peace Prize in 2004 for her contribution to sustainable development, democracy and peace and her fight to promote an ecologically viable development. The politics of representation from women in developing countries thus come a full circle, branching out into several networking activist websites and projects on climate justice.

Above we stated that a focal point for interdisciplinary inquiry into climate change might be lived experiences that themselves have no respect for the limiting knowledge imposed by independent disciplinary boundaries. The key question posed by the politics of representation, however, is: 'Whose lived experiences do we apply to interdisciplinary inquiry?' Without a concerted effort, it would not be the hungry of Kalahandi discussed above, or the waste pickers of Delhi (Box 4.3), or of others whose stories we have all-too briefly sketched in this chapter.

Nevertheless, we have also referenced the ways in which these stories have come to wider attention, and been forced onto the agendas of social scientists. To continue this process will help ground interdisciplinary inquiry in lived experience, not as some abstract notion of the concept, but as a broad spectrum of human realities and our various moral and ethical dilemmas. In this it will further help fulfil the interdisciplinary promise of delivering a broader and deeper representation of climate change.

This discussion concerning which voices are included or excluded actually relates to the specific politics concerning *whose* lived experiences of climate change are represented. In addition, however, the politics of the representation of lived experiences concerns the *ways* in which they are represented by different actors with diverse professional standpoints. Thus, in this chapter, the Indian examples of waste picking in Delhi and drought in Kalahandi are created by an NGO and a journalist respectively. In Chap. 1, moreover, we provided four main examples of lived experience:

- The UK floods experiences were represented by a journalist. In the full article, the journalist, Laura Barton, does present mostly the direct words of the interviewees. Nevertheless, we are obviously not provided the full transcripts and she will have edited the interviews. In this chapter there was then further significant editing and analysis by us, and hence further representation.
- The story of Joseph and Hurricane Katrina in New Orleans. This was represented by two social scientists, working within the broad field of social policy. The account contains a solitary, single-sentence quotation from Joseph. Otherwise it is completely in reported speech. Again, we in the chapter add a further layer of representation to edit further Joseph's story in order to bring out the salient points in relation to Hurricane Katrina.
- The Tove Valley Action group opposing the wind farm in the UK. Here we used the Action Group's website, which had a purpose in communicating to its own constituency the reasons for opposition, but also to local and national

politicians, to other groups opposing wind farms elsewhere and the general public. In this sense it was a representation of themselves for communication purposes, suitably edited for expression on a website. For our representation, we in turn have provided you the reader with an analysis based on an edited version of the key points made by the website.
- The accounts of two environmental activists, from Uganda and Zimbabwe. They were written in their own words, but their measured tone suggests that they know that their stories are going to be communicated to a wider audience and in a sense they are representing themselves. Then we add our comparative analysis!

In the above we witness representations by a journalist, a social scientist, a group of campaigning citizens, two environmental activists, and a further layer of representation by us. Are these eclectic sources of lived experience equally valid in the ways that they represent lived experience? Note also, that none claim explicitly to represent lived experience, and that interpretation is applied by ourselves to substantial parts of their accounts. To put it more bluntly, is the social scientific account more valid than those of the journalist, and even those of the campaigning citizens and environmental activists who are representing themselves?

Also, how do we (the authors of this book) assess the validity of our own analyses of these examples and others that you will find throughout the book? In general ways, and not now making a grand distinction between social and natural science, we have behaved like scientists in that:

- Our study of the lived experience of climate change involves searching for and establishing patterns from data that comprises accounts of people's lives.
- From these patterns, it proceeds to search for 'deeper' meaning and, if not universal principles, at least generalisations that apply within particular social circumstances.
- The soundness of our findings and their evolution are established through 'dialectics', that is, through critical engagement with others. While Aristotle may have suggested that this process must be distinguished from demonstration of causal processes through systematic inquiry, it is certainly part of the way that modern science proceeds in practice (see Chap. 3).

On the other hand we are obliged in our work to:

- Point to the absence of rigorous, controlled experimentation. This is impossible in our field where we have no control over the variables.
- Note that we embrace subjectivity, and make no attempt to reach the scientific norm of universal objectivity. We know that our findings are always provisional and subject to revision through the ongoing dialectics.

These two qualifiers place us firmly in the camp of qualitative social research with its critique of (i) empiricism that embraces controlled experimentation and (ii) the quest for objectivity in the natural sciences. Thus, we have no option but to work with untidy data, as Sect. 4.5 puts it, where pattern searching from different sources and interpretation through collective dialectics are our sources for establishing soundness. This process, however, is much more positive than implied because analysing such diverse sources also stimulates the mind to reveal fresh, creative insights.

All of these final issues and opportunities that we have raised in this chapter matter if lived experience is to be of any use in guiding policy and action on climate change. This is why issues of representation have been to the fore in Chap. 4. They cannot be ignored even if they do present fundamental difficulties. Following the Ancient Greek philosopher, Aristotle (see Chap. 3), insofar as we, the authors of this book, are scientists ourselves: we observe what is there, search for meaning in what we observe, and offer our observations and meaning to the world through representation.

We now turn in Chap. 5 to a particular way that climate change and the lived experience of it are represented in a significant literature—that of local and indigenous knowledge.

References

Abbott, D. (1995). *Methodological dilemmas of researching women's poverty in third world settings: Reflections on a study carried out in Bombay* (Open University Working Paper. 24, pp. 1–40). DPP. Faculty of Technology.

Berg, B. L. (2004). *Qualitative research methods for social sciences* (5th ed.). Boston: Pearson/Allyn and Bacon.

Boulton, G., Peter, C., David, E., & James, N. (2010). *The independent climate change emails review*. http://www.cce-review.org/pdf/final%20report.pdf. Accessed February 24, 2015).

Bulmer, M., & Warwick, D. (1993). *Social research in developing countries: Surveys and censuses in the Third World*. London: UCL Press.

Conklin, J. (2001). *Wicked problems and social complexity.* PDF CogNexus Institute: 11. http://cognexus.org. Accessed February 20, 2015.

Dee, D. P., Uppala, S. M, Simmons, A. J., et al. (2011). The ERA-Interim reanalysis: configuration and performance of the data assimilation system. *Quarterly Journal of the Royal Meteorological Society*, *137*(656), 553–597.

Conklin, J. (2005). *Wicked problems and social complexity in dialogue mapping: Building shared understanding of wicked problems* (pp. 1–25). Chichester: Wiley.

Famine early warnings system (FEWS). http://www.FEW.net. Accessed February 20, 2015.

IPCC, Flato, G., Marotzke, J., Abiodun, B., Braconnot, P., Chou, S. C., & Collins, W., et al. (2013). Evaluation of climate models (Chap. 9). In Stocker, T. F., Qin, D., Plattner, G.-K., Tignor, M., Allen, S. K. & Boschung, J., et al. (Eds.), *Working Group I contribution to the IPCC Fifth Assessment Report Climate Change 2013: The Physical Science Basis* (pp. 741–866). Cambridge University Press: Cambridge, New York. doi:10.1017/CBO9781107415324.020.

Fourier, J. (1824). *Memoirs of the Royal Academy of Sciences of the Institut de France* (pp. 585–587, cited in Bolin 2007 p. 3).

Pouillet, C. (1838). *Mémoire sur la chaleur solaire, sur les pouvoirs rayonnants et absorbants de l'air atmosphérique et sur la température de l'espace* (cited in Bolin 2007 p. 3).

Gupta, S. (2003). 18 years later: The poverty show goes on June 18. *The Times of India*. http://timesofindia.indiatimes.com/india/18-years-later-The-poverty-show-goes-on/articleshow/29969.cms. Accessed February 28, 2015.

Hickman, L., & Randerson, J. (2009, November 20). Climate sceptics claim leaked emails are evidence of collusion among scientists. *The Guardian*. http://www.guardian.co.uk/environment/2009/nov/20/climate-sceptics-hackers-leakedemails. Accessed March 2, 2015.

Hulme, M. (2009). *Why we disagree about climate change: Understanding controversy, inaction and opportunity*. Cambridge: Cambridge University Press.

Hulme, M. (2014). *Can science fix climate change?* Cambridge: Polity Press.

Miles, M. B., & Huberman, M. (1994). *Qualitative analysis: An expanded sourcebook* (2nd ed.). Thousand Oaks: Sage.

Lazarus, R. J. (2009). Super wicked problems and climate: Restraining the present to liberate the future. *Cornell Law Review* (Vol. 94, pp. 1153–1233). Also available: http://www.law.georgetown.edu/faculty/lazarus/docs/articles/Lazarus_SuperWickedProblem.pdf. Accessed January 12, 2015.

Rittel, H. W. J., & Webber, M. M. (1973). Dilemmas in the general theory of planning. *Policy Sciences, 4*, 155–169.

Roberts, H. (Ed.). (1981). *Doing feminist research*. London: Routledge and Keegan Paul.

Russell, B. (1931). *The scientific outlook* (1st ed.). London: George Allen and Unwin.

Sainath, P. (1996). *Everybody loves a good drought: Stories from India's poorest districts*. New Delhi: Penguin.

Skaburskis, A. (2008). The origin of 'wicked problems'. *Planning Theory and Practice, 9*(6), 277–280.

Tyndall, J. (1859). On the transmission of heat of different qualities through gases of different kinds. *Proceedings of the Royal Institution, 3*, 155–158.

Chapter 5
Lived Experience and the Advocates of Local Knowledge

Abstract Local knowledge is promoted by many as a key component of designing and implementing interventions on climate change, especially interventions that aim to adapt to changing climatic conditions. It is seen to result in interventions that are appropriate to and work in local contexts. It also demonstrates to policy makers that poor and vulnerable communities have innovative agency, and are thus empowered by their local knowledge. Nowhere are these facets better illustrated than in the annual Community-based Adaptation Conferences that are organised by the International Institute of Environment and Development. Challenges arise, however, when locals and outside professional experts attempt to share their different forms of knowledge to design and implement interventions on climate change. Formal participatory processes are intended to meet these challenges, but they are not immune from enduring power relations between insider and outsider and possible exploitive uses of local knowledge. Beyond practical intervention, local knowledge may be seen as linked to the identity of, especially, vulnerable groups in rural areas of poor countries. Here there are particular issues of passing local knowledge from one generation to the next. Finally, local knowledge is a manifestation of lived experience. Local knowledge may also, with care, be sometimes appropriate as a proxy indicator for lived experience, but should not be viewed as synonymous with it. Lived experience is not necessarily restricted to local circumstance and it extends beyond practical application in, say, climate change interventions.

5.1 Introduction

Issues of environment in relation to economic and social development of poor countries propelled themselves up the international political agenda in the 1980s, coinciding with burgeoning concern over global warming (Chap. 4).

Thus, the United Nations convened in 1983 a 'World Commission on Environment and Development' (WCED) chaired by the then prime minister of

D. Abbott and G. Wilson, *The Lived Experience of Climate Change*,
DOI 10.1007/978-3-319-17945-2_5

Norway, Gro Haarlem Brundtland. Commonly known as the Brundtland Report, the WCED published *Our Common Future* in 1987. There it combined environment and (social and economic) development within the concept of 'sustainable development', which it defined as:

> Development that meets the needs of the present without compromising the ability of future generations to meet their own needs. (Brundtland 1987: 43)

This famous definition is still widely in use today, despite numerous critical interrogations over the intervening years. *Our common future* certainly raised sustainable development as a global challenge. As a result of the report, and fuelled by rising concern over global warming, the first 'Earth Summit'—the United Nations Conference on Environment and Development (UNCED) was held in Rio de Janeiro, Brazil, in 1992, attended by 172 governments from all over the world, of which 116 were represented by heads of government.

It is not the purpose of this book to engage with the well-worn debates concerning the overall achievements or otherwise of that summit, but to note one particular legacy—how it connected the 'local' to the 'global'. In particular, a key influential document from the summit, titled 'Agenda 21', was accompanied by the slogan, 'Act local, think global', with a recurring message that 'sustainable development initiatives are likely to be most effective at a local scale involving ordinary people' (Winter 2009: 244).

Agenda 21 and its emphasis on local action prompted numerous initiatives world-wide during the decade that followed the UNCED summit. The approach was undoubtedly influenced by the concomitant shift during that period in the nature of international development aid to poor countries, from high-handed top-down approaches initiated by international agencies and national governments (approaches which were perceived to have resulted in much failure, and wasted effort and money) to bottom-up approaches that involved local beneficiaries and their grounded knowledge.

In relation to climate change, the most obvious representation of the local and its importance has been the idea of 'community-based adaptation' that is promoted by the International Institute of Environment and Development (IIED), and which organises a conference on the topic for activists and others that has been annual since 2009.

The first of these international conferences was held in Dhaka, Bangladesh in 2005. Then, in a briefing paper to the second conference in 2007, IIED defined the target audience and described the basis of the modus operandi for community-based adaptation:

> CBA begins by identifying the communities in the developing world that are most vulnerable to climate change. These are generally very poor, depend on natural resources and occupy areas already prone to shocks such as floods or droughts….
>
> When the cooperation of the local intermediary and the community has been obtained, the process of identifying what adaptations are appropriate can start. This requires initial learning about the community's indigenous capacities, knowledge and practices of how to cope with climate hazards in the past. New activities, technologies or practices can then be introduced. (Huq and Reid 2007).

In the above and many other statements, we see that the local is often valued highly by climate change practitioners and academics for its knowledge, sometimes also called 'indigenous' or 'traditional' knowledge (see below). Ayers and Forsyth (2009) provide a classic example in Box 5.1:

Box 5.1 Local Knowledge in the Arctic

Arctic communities have always lived in conditions of environmental fragility and high variability, and as a result, they have a wealth of traditional knowledge and cultural practices that are helping them to cope with and adapt to new challenges. For example, the Inuit community of the island of Igloolik in Nunavut, 320 km north of the Arctic Circle, has developed many behavioural responses to climatic exposures. Indigenous knowledge passed from generation to generation has resulted in flexible hunting and harvesting techniques about what to hunt, when, and where; and how to tell when dog teams indicate thin and unsafe ice is in certain areas. People have also developed risk-averse practices, such as taking extra supplies of food and gas on expeditions and never travelling alone or on land or water where the ice seems thin… .

A study on developing appropriate adaptation policy in Nunavut, Canada undertaken in 2006 recommended that interventions to reduce vulnerability in the Arctic region should be developed in partnership with the local community and draw on this important indigenous knowledge and existing local coping strategies.

Proponents of local knowledge point to its importance when designing climate change and other interventions. As with lived experience it is a different form of knowledge that may sit uneasily alongside scientific and expert forms. What, however, is the relationship of local knowledge to lived experience? It is an obvious question to ask, because at first glance they appear to be similar ideas, if not the same. A reviewer of one of our journal papers (Abbott and Wilson 2014) suggested bluntly that we had simply 'rebranded' local knowledge as lived experience.

The ultimate objective of this chapter is to argue that this is not the case. Lived experience is not simply an old wine in a new bottle. To make our argument we will first engage critically with local knowledge, the claims made on its behalf, its successes, but also its conceptual shortcomings that concern its predominately apolitical and technocratic usage as practice and its static, unproblematic relationship to culture.

Nevertheless, the reviewer's comment was understandable, especially as there are several variations around the label 'local knowledge'. For example, 'indigenous knowledge' and 'traditional knowledge' have already appeared in the chapter and seem to be interchangeable. 'Indigenous technical knowledge' is also widely used. However, while local knowledge is 'knowledge about local context or setting'

(Fischer 2000: 146) and concerns what works empirically in those contexts/settings, 'indigenous/traditional knowledge' adds a temporal dimension. It signifies knowledge that has evolved locally over a long period of time and has passed from generation to generation among a particular group of people, for example forest dwellers in Kenya and Uganda (see Chap. 1 and Sect. 5.3) or a fishing community on a Pacific island. It suggests a connection between local knowledge and way of life for a group of people, in other words their culture and their identity. This makes indigenous/traditional knowledge broader than the practical connotations that accompany local knowledge. It is perhaps for that very reason that others have reined it in by inserting the word 'technical'. Thus, like local knowledge, indigenous technical knowledge is rooted in practices of a group of people, while recognising explicitly that these practices have evolved locally over a long time. Examples (among many) might include practices of water conservation in India (Fig. 5.1) or selective breeding of drought and flood-resistant food grains in rural Africa.

It is also interesting to note that the idea of local knowledge and its variant appears to be more pertinent to people who live predominantly in rural areas in poor countries or are marginalised groups in rich countries (for example North American Indians). It is not unknown to use the term among affluent social groups, but in poor rural communities it is seen as being much more relevant at two levels—(i) to livelihoods and survival; (ii) to a way of life and identity. It is probably because of this that local knowledge has featured significantly in debates, policy and practice concerning development aid interventions, where invariably it is linked to local environments and where the weather in the form of rainy and dry seasons has always been a primary concern.

A further contribution to the debates, that may be obtained from development aid reports and the academic literature associated with the discipline of development studies, concerns the *process* of incorporating local knowledge into appropriate interventions. Much has been made in this literature of participatory practices, where professional expertise meets local knowledge to 'do development' together in partnership. In the idealised version of the participatory practice, everyone's input is equally valid and a positive process of dialectics takes place, out of which good interventions emerge where experts and locals alike have a strong sense of ownership. In this version, inclusive, direct, democratic practices have ensured that local knowledge is valued and been incorporated into good interventions.

Of course the real world is not like that, and the development studies literature has provided many refreshing critiques of participatory processes. We introduce these as the chapter progresses, but to give you a taster, albeit crude: Leroi, whose book about Aristotle and science featured strongly in Chap. 3, also had a view on the evolution of direct, participatory democracy in Athens, circa fourth century BC:

> Public life… was squalid. Every citizen could go up to the Pnyx and vote on the legislation of the day. Many did – if only for the sake of the three obols they got for attending. The result was institutionalised mob rule [where] demagogues roused the rabble (Leroi 2014: 316).

Nobody is suggesting that participatory development practices are like Leroi's description of ancient Athens, but, as in that extreme example, ever-present are the underlying power relations between actors that pervert the ideal.

Fig. 5.1 Resurrection of indigenous knowledge in Jaipur, India. This channels system reflects ancient knowledge of water conservation. The channels bring rainwater from the surrounding hills to Jaigarh Fort which was built circa fifteenth to eighteenth centuries. There they fill three underground tanks, capable of storing almost 30 million litres of water. The system is surprisingly still intact and so efficient that even today it is opened up in times of need to supplement the municipal supply of the city of Jaipur, for example during droughts and other times of increased demand. (*Photograph* Frank Abbott July 2014)

5.2 Local Knowledge and What Works

This section concerns the practice of local knowledge in adapting to climate change impacts. Generally speaking, the practical use of local knowledge is the predominant of the two ways in which it appears in the climate change, environment and development literatures, as identified above. We cover the use of local knowledge to affirm a group's identity—the second way—in Sect. 5.3.

Note the phrase 'adapting to climate change' in the paragraph above. Local action does not usually result in actions that attempt to *mitigate*, still less stop, climate change per se. To do so would require at least stabilising the concentration of carbon dioxide and other greenhouse gases in the atmosphere in order to halt global warming. While mitigation actions might be local, such as personal reduction in energy consumption (especially by richer consumers) and planting trees to absorb carbon dioxide, the knowledge that they rely on is the generalised science of anthropogenic warming and its sources. Local knowledge, on the other hand, tends to be used where current climate change is an experienced contemporary fact, the need to address its impacts is pressing and actions to adapt to the situation are therefore unavoidable. Mitigation and adaptation as concepts related to climate change are the focus of Chap. 6, which expands the definitions that are suggested here.

Thus local knowledge gives rise to local actions which usually comprise local adaptations to climate change. In rural areas of poor countries it is possible to find examples at almost every turn. In Sect. 5.1 we mentioned the (now) annual 'Community-based Adaptation Conference' whose venue is either a sub-Saharan African or an Asian country. It shares local practices and ideas, and ensures that innovations, which would otherwise be unseen or 'below the radar', are recorded and disseminated to a wider audience. Since its inauguration in 2005, the conference has recorded a rich seam of local innovations for adaptation. Box 5.2 gives some examples from the 2010 conference which was held in Tanzania, East Africa.

Box 5.2 Below the radar innovations in adapting to climate change from the Community-Based Adaptation Conference 2010.

The annual Community-based Adaptation Conference is organised by the International Institute for Environment and Development (IIED) based in London.

For two days prior to the 2010 conference in Dar Es Salaam (the capital city of Tanzania), participants undertook field trips to see practical examples of community-based adaptation. The following are taken from the official report of the conference, which was produced and published by the International Institute for Sustainable Development (IISD), a policy research institute based in Winnipeg, Canada, with offices in Ottawa (Canada), New York (United States) and Geneva (Switzerland).

Petra Bakewell-Stone, Pro-Natura International, offered highlights of a visit to the Kisarawe District, 50 km west of Dar Es Salaam, which included vulnerable communities affected by drought and water shortages. She said it remained to be seen whether the conditions were a result of climate change or mismanagement. In one of the coastal villages visited, the group witnessed a successful community project that was adapting to climatic changes by shifting to more sustainable mangrove harvesting after many agricultural sites had been submerged or abandoned.

Million Getnet, Haramaya University [Ethiopia] discussed his group's visit to three villages in the coastal Mukuranga region, which has seen an increase in temperatures and erratic rainfall in recent years resulting in drought and water shortages that have had a significant impact on maize and rice production. He said that to adapt to the situation, some of the villages have shifted to cassava production, a drought-resistant crop, as well as diversified into poultry production.

Nanki Kaur, IIED, told participants about a visit to a site in the Kinondoni Municipality, Dar es Salaam, close to the hotel venue, where increased rainfall intensity is having a significant impact on urban slums. She noted that communities have found innovative ways of coping with the impacts of climate change that do not require external support and huge investments, such as using rubbish as a flood defence.

Source: IISD (2010: 5).

For readers who might question the word 'innovation', note that participants in community-based adaptation invariably put their (local) knowledge to productive use with the materials that are available to create something new. That is a general definition of 'innovation' (Chataway 2005). Historically, the term evolved as a driver of capitalist economies, triggered by competition which forces private sector firms to do new things either in terms of their processes or in developing new products. This remains its major use. The innovations in Box 5.2, however, are triggered by the threat, already being experienced of climate change and are drivers of the process of adaptation.

Box 5.2 illustrates the diversity of local innovations in adapting to climate change. Local knowledge put to local use is nothing but contextual in terms of place. Also notable in these examples is the apparent absence of organisation and coordination at a larger scale. No government or international aid organisation is attempting to oversee and steer the processes of innovation. No scientific or other expert advice is being offered. The adaptations are driven by the collective needs of the communities themselves. More, these below-the-radar innovations are taking place irrespective of whatever adaptation is being attempted at larger scales (for example, large-scale flood defences). They are a far cry from the universal science of climate change and of attempts to reach international agreements through United Nations conferences. It is as if they exist in a parallel universe. This is what IIED Director, Camilla Toulmin, said in a closing speech to the 2014 conference:

> My own understanding of adaptation and reliance[1] draws very much from following over the past 35 years the process of change of a small village in Mali, West Africa... Over that period, they have been dealing with drought, managing scarce water, and growing different crops and diversifying livelihoods, and this has given me a fundamental respect for the knowledge, innovation and capacity that we find at field level in village communities across the world (Toulmin 2014).

A major role of the Community-based Adaptation conference, however, is to create an international community of communities that are adapting. It publicises their innovations, promotes sharing and supports the most vulnerable communities. As its organiser, the IIED situates it within a broad aim of 'Empowering Communities to Adapt to Climate Change'. The current IIED website describes the rationale as follows:

> Community-based adaptation to climate change focuses on empowering communities to use their own knowledge and decision-making processes to take action. (IIED, no date).

Recent conferences have made a concerted attempt to broaden the remit beyond sharing and supporting examples of already existing adaptations. This is reflected in themes of recent conferences. For example, the 2011 theme was 'Scaling up beyond pilots', by which it meant 'focusing on the need to spread CBA knowledge and practical lessons horizontally across communities and vertically across levels of governance and action' (IISD 2011: 1). In 2013 the theme was 'Mainstreaming into Government' by which it meant institutionalising community-based adaptation into international, national and local planning and processes' (IIED 2013). The 2014 conference in Kathmandu, Nepal, launched a declaration—'The Kathmandu Declaration on Financing Local Adaptation'—which called for strengthening national and international finance for local adaptation to climate change, and for steps to ensure that it reaches the most disadvantaged communities (IIED 2014).

The Kathmandu Declaration was significant because, once the subject is raised of outsiders financing local adaptations, the rules of the game inevitably become modified away from simply observing, publicising and sharing what is already happening without such finance. Whether this money comes from the private sector, national governments or international aid organisations, they will want to *know how* it is spent and to a significant extent will want to *determine how* it is spent. While 'upscaling' and 'mainstreaming into government' both inevitably require engagement with wider groups of stakeholders, involving some loss of power to self-direct local innovations, the balance is tipped even further if the other stakeholders are bringing finance to the negotiating table.

---

[1]Authors' note, we suspect that this is a typographical error in the transcript and should read 'resilience'.

It is for reasons such as these that participatory processes between 'insider' and 'outsider' have been proposed and developed to address the challenges of multi-stakeholder engagement involving local and expert knowledges. Ayers and Forsyth (2009) state in relation to community-based action on climate change:

> A community-based approach considers that adaptation strategies must be generated through participatory processes, involving local stake-holders and development and disaster risk–reduction practitioners, rather than being restricted to impacts-based scientific inputs alone. As such, expertise in vulnerability reduction must come from local community-based case studies and indigenous knowledge of locally appropriate solutions to climatic variability and extremes.... From a community-based adaptation standpoint, climatic changes are acknowledged by trying to understand locally observed, historic patterns of change in ways that are meaningful to local people.

5.2.1 Sharing or Extracting Knowledge Through Participatory Processes: The Contribution of Development Studies

The principle of participation is therefore simple: both local and external stakeholders engage and share their knowledge and agree what to do in, for example a climate change intervention, and within the constraints of the budget.

Thanks to Development Studies, however, we have a wealth of analysis and debate about participatory processes, from firm advocates to those who have considered it a 'new tyranny'. The champion *par excellence* has been Robert Chambers who argues that too many development aid interventions are based on positivist, exclusive, expert knowledge, where professional development experts sustain 'their own reality that they transfer to others (Chambers 1997: 54). He has insisted, therefore, that outsider experts should 'hand over the stick' and listen to the knowledge of local 'beneficiaries' of development interventions (Ibid.). This will lead to better interventions and local ownership through a democratic process of deliberation (Ibid: 188). These arguments were later mainstreamed by the large international development institutions such as the World Bank, which produced three books on the overall theme of 'Voices of the poor' that took on board and to a large extent institutionalised 'participation' within its own processes (Narayan and Petesch 2002; Narayan et al. 2000a, b).

In contrast to the enthusiasm of Chambers and others, the 'tyranny' critique focuses on the failure of participation to address issues of power and politics, issues which pervade all spaces where international development interventions are created. This critique was brought together in an edited collection by Cooke and Kothari (2001). Of direct interest to our book, other authors of that period focused on the knowledge dimension of the tyranny, suggesting that local knowledge was extracted by development experts during participatory processes. To be fair, Chambers (1997: 111, 214) had previously suggested that this is a potential danger but the later authors were more strident. For example, writing about urban planning, Rakodi (2000: 21) suggested that the participatory methods applied to town planning and decision making had resulted in little more than 'more aware and better informed "outsiders"'. Meanwhile the World Bank and International Monetary Fund were demanding that recipient poor-country governments of their financial loans produce Poverty Reduction Strategy Papers (PRSPs) that involved participatory processes. A review by Craig and Porter (2003: 54) described them as a 'Third way for the Third World' where knowing the population is crucial for success. They elaborated by stating that PRSPs are supported by 'enhanced information and statistical methods for identifying, mapping, measuring and reporting on poverty, as popularized by publications such as "Voices of the Poor"'.

One of the authors of this book, Wilson (2006), reviewed these debates with respect to a common pejorative label of development interventions at that time. This was that they are 'technocratic', in other words defined, designed and implemented by experts, or 'knowledge elites', which stripped the interventions of all politics and refused alternative ways of thinking about the problem. Chambers had been one of the leaders of the charge on this, strongly criticising mainstream development interventions for being bad, top-down and undemocratic. Wilson's review, however, suggested that a transition was taking place from the classical definition of technocrats as 'knowledge elites' in international development to a new kind of technocrat, 'learning elites', of the early part of this century whose job it is to extract and assimilate local knowledge. To perform the role of learning elite, however, requires a methodology that is different from the usual census-style statistics gathering. Participation fitted the bill exactly and became new methodology of choice (Wilson 2006).

Wilson's critique drew on Foucault's (1979) notion of 'governmentality'—the art of modern government—which requires knowledge of the population to be governed, with the aim of enhancing its productive potential. In Foucault's own words: 'The population is the object that government must take into account in all its observation and "savoir", in order to be able to govern in a rational and conscious manner. The formation of a "savoir" proper to government is absolutely bound up with the knowledge of all the processes related to population' (Ibid. 18).

Extracting knowledge for the learning elites was, in Wilson's view, a prime example of this Foucauldian notion of governmentality. However, while the governmentality argument is a neat formulation of what happens to local knowledge during participatory processes, it is too neat. Yes, power relations exist between the different actors, but the very fact that outside experts might wish to extract local knowledge for purposes of governmentality provides some leverage and power to the holders of that local knowledge. As such, participatory processes may have all kinds of unanticipated effects (Cornwall and Andrea 2004: 85) where the potential exists for 'genuine negotiation' between the locals and other actors (Leach et al. 2005: 217). Foucault's notion of governmentality and Cornwall's analysis of participatory processes are further analysed in Chap. 10.

It is in this vein of deepening and moving beyond knowledge extraction and tyranny analyses that Sam Hickey and Mohan (2004) edited a follow-up book to the original collection by Cooke and Kothari. This book examines the positive possibilities of participation alongside the tyranny argument. Such possibilities include using participation to engage with a 'radical political project on the basis of promoting citizenship' and, where it cannot claim to be transformative, 'to find a new "gradualist" language to extol its virtues' (Hickey and Mohan 2004: 20).

5.2.2 Local Knowledge, Power and Participatory Processes in Environmental Studies

The above debates that abound in development studies are also present in the environmental literature, although they tend not to go into the same depth. One reason for the comparative neglect is, perhaps, the strength of the underlying idea in environmental discourse that we all must pull together to safeguard our environment, including our climate, otherwise the consequences for life on earth could be dire. Thus, Frank Fischer who wrote a book substantively about local knowledge, experts, participatory processes and the environment (Fischer 2000) echoes much of Chambers' (1997) work in relation to development studies. Fischer notes the importance of local contextual knowledge to challenge the scientific emphasis on "generalisable knowledge"; claims that local knowledge is the primary product of participation; and that 'ordinary local knowledge' is important for 'problem identification, definition and legitimation' (Fischer 2000: xii, xiii, 217). These comments that concern integrating local knowledge into mainstream deliberation about the environment, however, could easily have been written by Chambers. The same could be said of Fischer's 'solution', which is to reconstruct professional practices so that experts are also facilitators who listen to what the locals have to say and to accept their realities. Again like Chambers, he recognises the 'social and political' problems of enacting this solution, offering that 'much will depend on a commitment to participation on the part of both professionals and society as a whole' (Ibid: 191, 192). Asking for such voluntaristic 'commitment', however, appears to evade the crucial issues of power relations, and hence politics that forms a central tenet of Fischer's book.

Finally, while noting the possibilities for integrating local and expert knowledge through participatory processes, Fischer also alludes to the extractive potential when he states that 'ordinary citizens… can also contribute a form of knowledge—local knowledge—that the professional expert requires' (Ibid: xii). While critiques of participation in development studies are pejorative about the extractive potential, this citation suggests that Fischer is at least neutral and possibly positive about it.

The expert-local relationship also forms the object of study in what is known as the environmental adaptive and co-management literatures. Adaptive management is conceived as a continual process of social learning (Box 5.3) among diverse stakeholders, while co-management is concerned directly with user participation (Armitage et al. 2007: 1). There has been at least one attempt to bring the two together as 'adaptive co-management' (Ibid.)

Box 5.3 Social Learning as used in this book

Much has been written about social learning, but for our purposes in this book it simply means learning through interaction with others in a particular social context.

Few articles or books in these genres, however, make more than passing references to the differential power of diverse knowledges. For example, Williams and Brown (2014) offer an adaptive environmental management framework that emphasises the 'importance of stakeholder involvement …, both initially in the design of the management framework and throughout the iterative process'. Such involvement is essential of course for the underpinning social learning process and the integration of stakeholder knowledges, yet the framework is presented as a set of neutral, apolitical procedures. The word 'power', meaning social power attached to different forms of stakeholder knowledge, is totally absent.

In November 2013, the journal 'Environmental Management' published a special issue comprising a selection of papers under the theme 'Indigenous Studies Speaks to Environmental Management'. In the introductory paper, the special issue authors explained their express intention of examining 'some of the ways in which an Indigenous studies perspective can guide thinking about environmental management, or environmental stewardship'. They further expressed the view that the concept of 'stewardship' may be more familiar to indigenous groups that have been 'involved in the "management" of environmental and natural resources on their lands since time immemorial' and who have particular forms of knowledge and worldviews to bring to the topic (Richmond et al. 2013).

In their preview of the rest of the special issue the authors (Ibid.) noted: 'Several papers highlighted the power differentials that emerge when western environmental management practices require Indigenous groups to convert their different forms of knowledge and practice into a western format that environmental managers can utilize'. This is true. Some papers do indeed highlight issues of power differentials related to knowledge. None, however, stretch beyond noting that they constitute a major challenge to be overcome and providing examples. To illustrate this point, here are two illustrations from the special issue that relate directly to indigenous knowledge:

(i) To develop methodological practices that accord equal power to differing peoples and knowledge systems, this requires that both western and Indigenous knowledge systems be equal in the "co-production" of regulatory data. This would bring greater democracy to the sciences— instead of bringing science to democracy. Dealing in these ways with power relationships might lead to learning about the social–ecological system in ways that adaptive co-management initiatives so far only hope (Watson 2013).

(ii) We must begin to conceptualize conservation beyond "environment versus human activity" to encompass a respectful, multigenerational form of human–environment

> interaction embodied in the recognition of indigenous environmental knowledge and practice. (Middleton 2013, writing about environmental legislation and protection of Indigenous Rights among United States native Indian groups)

5.2.3 Local Knowledge and the Climate Change Literature

The insights from development studies, and to a lesser extent environmental studies, about participatory processes and power relations between local and expert knowledges have not alas been replicated to any extent in the climate change literature. Overtly political power struggles between stakeholders are covered, however, as do prevailing capitalist national and global power structures which often hinder meaningful action on climate change. See for example the edited collection of Carrapatoso and Kürzinger (2014), and we take up some of the issues in Chaps. 9–11 on lived experience, public action and policy.

The climate change literature does sometimes refer to the power differentials between local and expert knowledge in an oblique way. For example, Derbile (2013) analyses how farmers in the Atankwidi basin, north-eastern Ghana, are reducing vulnerability of rain-fed agriculture to drought (that he links to climate change) through 'indigenous knowledge systems (IKS)'. He finds that 'farmers are planting multiple indigenous drought resilient crop varieties and employing different rounds of seeding and/or staggering planting between multiple farms. They are also applying indigenous forms of organic manure, checking soil erosion through grass strips and stone terracing and adopting paddy farming for improving soil and water conservation towards enhancing plant adaptation to drought.'

As a result of this research Derbile (Ibid.) recommends 'giving priority to the strategic role of IKS in climate change adaptation planning'. So far so good, but he then further claims that IKS is a 'science', increasingly recognised as such, but lags behind mainstream 'western' science only through lack of documentation and the continuing dominance—that is to say, power—of the latter. Thus for Derbile (among others who make similar arguments), the task is to construct IKS as a science that will have equal standing with 'mainstream, western science'. In other words, this will be the means by which the power differential is removed.

Is this, however a valid construction? In terms of criteria for being counted as a science, have the Atankwidi basin farmers conducted a systematic investigation? Have they separated the problem into its constituent elements when thinking about it? They have almost certainly experimented, but have they done carefully controlled experiments that have compared a deliberately manipulated situation to an un-manipulated control? (see Chaps. 3 and 4 of this book). The point is that the Atankwidi basin farmers have done something much richer than conduct a scientific investigation. They have marshalled holistically knowledge gained over a long period of time to develop seeds and practices to meet the contemporary problem of drought. This knowledge can stand on its own. To label it as 'science' is actually to accept the 'superiority' of what usually goes on under that label and this, in our view, is counter-productive.

In the practice-based literature on climate change, examples are given of fruitful engagement between local and expert knowledges, where power relations are never mentioned. Box 5.4 concerns local engagement in Sudan. It is taken from a report on integrating climate change adaptation into secure livelihoods, produced by the international development NGO, Christian Aid (reported in IIED 2009b: 145). It describes knowledge engagement between representatives of the local community in Atbara and meteorologists, a town in Sudan where the River Atbara joins the Nile.

Box 5.4 Developing a climate timeline in Sudan

Atbara Partners Consortium members developed a climate timeline as part of a workshop to review a climate change awareness-raising project. The members determined the main climate features that affected their lives and livelihoods and discussed the trend over the past 30 years, but with a focus on the last 10 years as this was (a) easier to remember and (b) a time in which there was agreement that changes away from the normal variation had occurred. Also involved in the discussion were two staff from Atbara Meteorological Station, who questioned consortium members' perceptions, corrected dates, and gave statistical evidence from their records.

The notable disagreements centred on rainfall patterns, with meteorologists conceding that with limited rain stations and the localised nature of flash floods, these could be missed in the records. Increased average temperatures were not verified by the statistics but members' perceptions could be linked to increased humidity (which feels hotter) and increased variation (hence the 2007 record high), which was suggested as an emerging pattern by the climate scientists. The challenges agreed by the consortium are to:

- deepen the analysis to include differences in trend and variability;
- continue the process to determine what the likely scenario for the next 10 years will be;
- how this will affect the vulnerability of their livelihoods; and
- what therefore the consortium should do in terms of project development and implementation.

(Source: IIED 2009).

Box 5.4 offers a useful end note to this section. It illustrates that one may set up participatory processes and frameworks for engagement between expert and local knowledges, but in the final analysis much depends on the actors involved—their willingness to share and the incentives for them to do so. In this case, the meteorologists obtained knowledge about localised weather patterns that they did not previously have. In the words of Frank Fischer quoted above, the local community possessed knowledge that the professionals required. The community in return obtained scientific information about recent weather patterns that they could correlate with their own experiences. What we see in Box 5.4 is an example of apparently successful sharing because of a high degree of mutual gain in knowledge from each other. Elsewhere, Johnson and Wilson (2006) report on the importance of mutuality for successful knowledge sharing, in this case between Ugandan and British environmental health officers.

5.3 Local/Indigenous Knowledge and Identity

Beyond 'practical' local knowledge, there is also a sense in which it may be associated with the identity of a social group. This is especially so when 'local' is conceived as 'indigenous', indicating knowledge that has evolved over generations. As well as its practical dimensions, such knowledge is then seen as expressing a group's culture in terms of its way of life and particular values of what is right and wrong, and with ways of making sense of our ontological selves. The last represents our fundamental ways of being which are often expressed as our 'world views', from the German word 'Weltanschauung'.

We referred in Sect. 5.2.2 to a special issue of the academic journal 'Environmental Management'. In the introductory paper, Richmond et al. (2013) touched on the potential clashes between diverse world views in environmental management processes that include indigenous interests. They pointed particularly to the challenges of reconciling indigenous and western management worldviews, and the possible subjection of indigenous world views when Indigenous groups are required to convert their different forms of knowledge and practice into a western format that environmental managers can utilise.

One could easily substitute 'identity' for 'world view' in the above paragraph. In fact, culture expressed as our ways of life and values, and our diverse ontologies expressed as world views, overlap with, and to a large extent may substitute for, 'identity' (we expand this notion of culture in Chap. 7).

In Chap. 1 we analysed a case study taken from Elvin Nyukuri's Ph.D. thesis (Nyukuri 2013), concerning the ways in which climate change policy is exacerbating the vulnerability of the former forest-dwelling Batwa and Ogiek tribes of East Africa. At three points in the thesis she comments on the role of indigenous knowledge in maintaining culture and identity, and as a means of exerting Batwa and Ogiek agency with respect to climate change mitigation and adaptation strategies when aligned with other indigenous groups:

> Indigenous knowledge is one area the Ogiek and Batwa have used to identify themselves [and exert their agency] …
>
> Because of this knowledge of the forest it helps the Batwa and Ogiek speak in one voice. Because of this knowledge the Batwa and Ogiek associate themselves with other indigenous groups internationally through institutions such as the Minority Rights Group [an

international Human Rights-focussed NGO with around 130 partners in over 60 countries—MRG: no date] because they have common knowledge that is beneficial in terms of mitigation and adaptation strategies. In addition, this is one area which the Batwa and Ogiek have strategically used to conserve their culture' (Ibid: 246, 257, 258).

Nyukuri also notes, however the ways in which this knowledge is being eroded and lost, partly as a result of the younger generation moving away to urban areas. She had previously commented that this is a general pattern among forest communities which weakens the performance of rites of passage (here, rituals that mark the passage into adulthood):

> [I] observed that the delay in performing rites of passage strained the capacity of sharing and the passing on of indigenous knowledge among the Ogiek (Ibid. 207).

> I noted how the loss of such [indigenous] knowledge was affecting the health of the Batwa. It was also evident that the elders could no longer pass on knowledge on traditional forest weather observations to the young generation. This was because of lack of access to the forest and also because they did not remain in one area for sufficiently long periods to observe such changes. (Ibid: 246).

In these statements, the causes and effects appear to be circular. The communities are already highly vulnerable and to a large extent have ceased to be forest dwellers in any substantive sense. Youth are moving to the cities in search of employment. As a result, identity as Ogiek or Batwa is being eroded and with it loss of indigenous knowledge which further serves to erode the identity of these tribes and increase their vulnerability in the world.

From Nyukuri's account, it seems that where indigenous knowledge has some practical value to decision makers on climate change, or perhaps is made to have some value through alliances of indigenous groups, it may also act as identity-reinforcing. Its erosion, however, contributes to loss of identity which feeds into a vicious circle of vulnerability, including, as Turner and Robbins (2008: 4) argue, loss of self-determination and influence. In other words, if knowledge is power, the converse is also true. Loss of knowledge is loss of power, especially if that knowledge is unique to your social group.

We end this section with a short case study (Box 5.5) that illustrates graphically the value of local practical knowledge and how the loss of this knowledge is both result and cause of a changing way of life in an apparently vicious circle. As with the participants in Nyukuri's case studies, changing attitudes of youth are central. The setting is an interview on the Pacific island of Vanuatu. The interviewer is talking to participants A and B about the impacts of tropical cyclones (hariken), whilst they work in B's Garden.

Box 5.5 Cyclones in the Pacific Islands

Interviewer So when the hariken came in 1939 you must have been 10—do you remember it?
A: Yes! ... Every tree went down... There was a white man that had a small store on Ra Island where my father worked and that day I went with him—and the big wind comes now! It came; it came until it pulled off the roof

belonging to the white man… the sea carried everything from the store right up into the middle of the island! We went and dug out tinned fish, soap—all things belonging to the store.

B: [It was the] worst hariken—we can't remember a worse one.

Interviewer: You had a hariken this year—can you tell me about that one?

B: [We have a] food shortage now! Oh yes, first time is this year. Small, small hariken but…

A: Plenty of hariken have hit us before but we have not had a food shortage. But this year—we have a shortage!

Interviewer: So in 1939 do you remember a shortage?

A: Small, small shortage. But all the old people then, they had good gardens and they stored plenty of dried breadfruit…

B: In 1972 it was the same [when a Hariken came]. The gardens were strong.

Interviewer: So what's different now?

B: I can't tell you straight—but I think it's because of a lazy fashion now! Oh, yes, they'll say they don't have enough land now, but the real reason is they don't want to work. There is enough land. We must plant something every day to make sure we have no shortage of anything—that was the fashion of the people before…

A: Custom! Custom belonging to us… must plant banana, taro, cabbage or whatever—every day.

B: That was the teaching belonging to our grandparents, that was the talk we used to hear in the Nakamal [meeting house], that was the talk we used to hear in the gardens with our parents. That was the talk before—before [the] school came. Plant plenty, plant a strong garden, then if disaster comes, you have food.

A … [We are] losing custom, that's why it happens: losing the custom/fashion belonging to the old people before.

(Source: International Institute for Environment & Development [IIED] 2009: 83).

5.4 Local Knowledge, Indigenous Knowledge and Lived Experience

The concept of lived experience has much to learn from previous work on local and indigenous knowledge, their relationship with expert knowledge and attempts to realise the potential to create an integrated knowledge through participatory processes. What, however, is the relationship of local/indigenous knowledge to lived experience? Put another way, are we simply rebranding local/indigenous knowledge as lived experience?

These questions that we asked in Sect. 5.1 formed the starting point for this chapter and now, having explored ideas that surround local/indigenous knowledge, we try to answer them.

Any common-sense view would suggest a close connection. We concur for, in our view, local and indigenous knowledges are *manifestations* of lived experience. In some circumstances they might become the major manifestations of lived experience, for example when our everyday lives are dominated by a local sense of place and/or a sense of identity of being part of a marginalised 'indigenous' group.

For the large majority of people on this planet, however, local and indigenous knowledges can never *be* lived experience, whether as old brand or new. This is for three reasons:

1. The lives we lead take place in multiple contexts, of which local place is one such context that has greater or lesser importance depending on the others. As outlined when we first conceptualised lived experience in Chap. 1, particularly important is the socio-economic context of our lives – whether we are men or women, rich or poor, high caste or low caste, and so on – and the livelihood opportunities that this context supports. Also, the related cultural context – expressed through the values of our particular social group - is important. This is not to argue that these other contexts are independent of local place. Indeed they are strongly connected, for if one is born poor, one is likely to live in a poor neighbourhood or a poor rural community. If one is born a woman, the home takes on added importance in most societies. We must guard, however, against reducing these multiple contexts to a single one. Lived experience recognises and accommodates the influence of multiple, inter-related contexts. Local context by itself, however, is local context *tout court.*
2. Our agency, which potentially enables us to go beyond the multiple contexts that shape our lives. We have a human ability to reflect on our own actions, on our communication and engagement with one another, and on the communications from others that we receive via the media. Communication and engagement confront us with the knowledge and lived experiences of others, and our reflection on these potentially enables us to expand our own knowledge beyond the boundaries of our particular contexts. Of course we must qualify what we are enabled to do through our agency that is not in circumstances of our choosing. Thus, it is true that if we are poor and our direct engagement with others is limited to other poor people within our locality, any knowledge expansion is likely to be limited. If we are poor women talking only to other poor women, knowledge expansion might be even more limited. If we are attached strongly to a particular social or religious identity, and only engage with those who are similarly attached, knowledge expansion beyond those boundaries is further limited. It is also true that we do tend to prefer engagement with those of similar mindsets—we do not necessarily wish to have our knowledge challenged at every moment. Despite all of this, most of us, however, have at least some engagement with those of different mindsets. Increased travel opportunities and, of course, the advent and take-off of digital social media facilitate such knowledge expansion. In turn, knowledge expansion through communication and engagement, feeds into our lived experience and it evolves. Chapter 8 explores these issues more deeply.

3. Lived experience creates its own rationalities that explain why we act as we do. These are subjective rationalities, rather than the abstract 'objective' rationalities that arise from scientific inquiry. In contrast, the rationality of most local knowledge abstracted from lived experience is simply whether it results in actions that work or not. Thus, changing the traditional food grain that one eats to one that is better suited to growing in a changing climate is considered objectively rational. Setting some land aside for growing the old grain on the other hand is considered irrational. Subjectively, however, the old grain might be considered superior in many other ways and continuing to use it is quite rational within one's lived experience. Recently, this subjectivity with respect to climate change has been labelled 'lay rationalities of climate change'. It has formed the subject of an academic journal special issue, focusing on 'lay knowledges that integrate individual understandings, explanations, life experiences, perceptions and intervention on climate change' (Alves et al. 2014).

In short, the above three reasons make clear that lived experience is forever evolving and is therefore dynamic. It is informed by our agency—our ability to reflect, communicate and engage—and by multiple contextual influences. It is certainly not limited to our knowledge of a place or as a member of an indigenous social group and cannot be reduced to these categories. It moves beyond the boundaries of the local and/or the indigenous group. In our view, because it is rooted in social process, it is more robust conceptually, and offers deeper insights into the nature of problems that are associated with climate change.

Thus, local and indigenous knowledges are manifestations, even outcomes, of lived experience. They are grounded in lived experience and may sometimes be used as a proxy representation of it, as you the reader may have already noticed in some of the case studies and examples that we have used throughout the book. As with all proxies, however, there are issues involved. Local knowledge is no better (or worse) a proxy for lived experience than the annual per capita gross income of countries (GNI) is of development, or infant mortality rate (IMR) is of health, or adult literacy level (AL) is of education. We miss a good deal of the richness, depth and complexity of development, health, education and lived experience if we reduce them to simple proxy categories that we can capture more easily than the complete phenomenon.

Sometimes it is inevitable that we use proxies instead of the real thing, however, if only to get us started on defining the issues. This is because IMR, GNI per capita and adult literacy may be measured and a number attached. Health as total well-being, development as 'good change' and overall educational attainment can't. Local knowledge, further reduced to its practical dimensions is more tricky, but may be captured through interviews and observations of local innovations. Lived experience in its totality, like health, development and education, however, is far too complex to capture easily, as we indicated in Chaps. 2 and 4, and it is certainly much more than practice.

The issue for researchers might be academic anyway, as the locals are unlikely to recognise the boundaries that experts put around their knowledge. In other words, local knowledge might be what one wants to capture but lived experience is what one gets, and it is then left to the researcher to extract what she or he wants.

Because local knowledge *is* embedded in lived experience, one may, however, extrapolate to the latter many of the opportunities and challenges associated with the former that we discussed in Sect. 5.2.

Thus opportunities of local knowledge/lived experience include:

- Demonstration that people are not passive in the face of climate change
- Demonstration of what works on the ground
- Knowledge that professionals require
- A more democratic way of making decisions through participatory processes

Challenges include:

- Acceptance of local knowledge/lived experience by scientists, other experts and decision-makers as a valid form of knowledge and the power relations between the different forms
- Possible exploitation where local knowledge/lived experience is simply extracted by experts and decision-makers and there is no mutual sharing
- The power relations between experts and lay people in participatory processes that are intended to engender space for sharing knowledge.

Writing about practical local knowledge, Fischer (2000: 217) adds a further opportunity, which is that integrating local and expert knowledges enhances the legitimacy of interventions in the eyes of the general public. We agree but also believe that it is important not to try and extend the claim by according local knowledge a power to make or break interventions that it does not possess. For example, when communities in western Europe and North America refuse to accord legitimacy to, for example, the establishment of a wind energy farm, they draw on much more than local knowledge, although that plays a part. They rationalise their opposition in many other ways, including marshalling scientific knowledge about the working of wind turbines and the damage to wildlife that might occur. Beyond that, they draw on their lived experience to identify the changes that the wind farm is likely to bring to their everyday lives, and indeed to their whole way of life. In short, it is the baggage in which local knowledge is embedded that accords legitimacy or otherwise to climate change interventions. We call that baggage 'lived experience'. We explore the part it plays in legitimising interventions on climate change in subsequent chapters, starting with Chap. 6.

References

Abbott, D., & Wilson, G. (2014). Climate change: Lived experience, policy and public action. *International Journal of Climate Change Strategies and Management, 6*(1), 5–18.

Alves, F., Caeiro, S., & Azeiteiro, U. M. (2014). Guest editorial: Special issue on 'lay rationalities of climate change'. *International Journal of Climate Change Strategies and Management, 6*, 1.

Chapter 6
Lived Experience and Discourses of Mitigation, Adaptation

Abstract Mitigation of, and adaptation to, climate change have been major themes of United Nations Intergovernmental Panel on Climate Change (IPCC) periodic reports since their inception. A review of the 2007 and 2014 reports, however, finds that the IPCC coverage of mitigation and adaptation, while increasingly extensive and drawing on both the social and natural sciences, restricts itself largely to a discourse that is driven by the physical phenomenon of climate change. There is no discussion, for example, of how the lifestyle and behavioural changes that it advocates in these reports may be achieved, and no substantive related discussion of the difficult political processes that inevitably underlie all intervention and which are crucial for their popular legitimacy. A lived experience lens inherently draws attention to these issues and to the current dominant underlying framing of climate change as a threat that requires a war and sacrifice in the present for future prosperity.

6.1 Introduction

Intervention on climate change is commonly classified under two broad headings—adaptation and mitigation. We provide the current United Nations Intergovernmental Panel on Climate Change (IPCC) definitions of these in Sect. 6.4 below, but simply, adaptation means adapting to the fact of climate change through, for example, better weather forecasting, building flood defences large and small, developing drought resistant or conversely flood resistant seeds, creating houses that float in flood-prone areas, relocating people who live in vulnerable areas, and so on (the list of potential adaptation actions is endless).

Mitigation means attempting to stabilise global temperature rise which is the underling physical driver of climate change. It involves a range of measures that are designed to reduce emissions of carbon dioxide and other greenhouse gases that arise from human activities, and interventions that 'capture' and remove carbon dioxide from the atmosphere, for example forests (and green plants generally) and carbon storage technologies.

D. Abbott and G. Wilson, *The Lived Experience of Climate Change*,
DOI 10.1007/978-3-319-17945-2_6

Adaptation and mitigation occupy a prominent position in the deliberations of the IPCC and in this chapter we explore what they have come to mean in this influential body. We find that, while the IPCC covers exhaustively a range of options for intervention and the challenges that they face, it glosses over others. One intervention it identifies in relation to adaptation and mitigation, but adds little else, concerns lifestyle change. Meanwhile, it fails to identify explicitly the challenge of popular legitimacy of intervention which we raised at the end of Chap. 5.

For this book, however, lifestyle changes and popular legitimacy are fundamental matters to address and they run throughout the sections below. They are fundamental because they raise significant challenges concerning the political nature of adaptation and mitigation decisions, something which the IPCC largely ignores. Although the IPCC Working Group (WG) III reports on mitigation do not use the actual phrase—'there is no alternative to taking strong action on climate change'—that is the sentiment that appears to be lurking in the background where the only permissible argument concerns, not political processes, but a limited range of 'strong' technical options. To be sure, the IPCC reports also refer to 'governance' arrangements, but they are usually aimed only at poor countries and are devoid of content. Taken as a whole, the reports state: the risks of not doing anything, what is possible to be done and what needs to be done. If politics plays a part, it can only be in a hidden assumption that governments should deliver, and do so in concert with other governments.

Conceived as political processes, however, attempts to engineer lifestyle changes may engender as much opposition as they garner support, while popular legitimacy may be sadly lacking for many interventions. A lived experience lens is a useful analytical tool for making sense of these issues.

6.2 Mitigation and Adaptation and the IPCC

The first IPCC Assessment Report was published in 1990, two years after the Panel was established in conjunction with the World Meteorological Office in response to increasing scientific and political concern at the international level (see also Chap. 1). Three working groups (WGs) contributed to the first Report: WGI Scientific Assessment, WGII Impact Assessment, and WGIII Response Strategies. Mitigation and adaptation as strategies for intervention were cited in the WGII report but only in passing and with no definitions or substance. As would be expected, they appeared with more frequency in the WGIII report on responses, including a whole section on 'Measures for adapting to global climate change'. Again, however there were no definitions.

The second IPCC Assessment Report was published in 1995, where WGII had an explicit brief to address both mitigation and adaptation. The title of WGII's report—'Impacts, Adaptations and Mitigation of Climate Change: Scientific-Technical Analyses'—made clear, however, that it did not cover the societal and political issues of interventions. In the same Assessment Report, WG1's title was slightly rephrased from 1990 to 'The Science of Climate Change' while that of WGIII promised for the first time an explicitly social science focus: 'Economic and Social Dimensions of Climate Change'.

The third IPCC Assessment Report was published in 2000, where adaptation and mitigation were now divided between WGII 'Impacts, Adaptation and Vulnerability' (the last word in this title suggesting a greater social science focus than hithertoo) and WGIII 'Mitigation'. Working Group 1's title was a further small variation on those used previously: 'The scientific basis'.

These working groups, with slight variations in title, have been entrenched in the 4th and 5th Assessment Reports that appeared respectively in 2007 and 2014. One title variation has concerned WGI which has made clear that it is not dealing with *anything* that might pass under the label 'science', but explicitly with the 'physical science'. You will see the exact WGII and WGIII titles as we cover them in more detail below, but for the moment note that from 2007 WGII has always featured adaptation and WGIII mitigation.

You should also note that IPCC reports have tried to deal with the issue of uncertainty in their findings, an issue which is inevitable given the large amount of evidence that they review and the scrutiny that it receives. The most systematic approach across all its working groups appears in the 2014 report, from a guidance note produced in 2010 on which our summary in Box 6.1 is based.

Box 6.1 Dealing with uncertainty in IPCC findings

The IPCC 5th Assessment Report relies on two metrics for communicating certainty in a finding: confidence in its validity of and a quantified measure of its uncertainty (expressed as likelihood of occurrence).

Confidence in validity

Each finding is grounded in an evaluation of underlying evidence (an amalgam of type, amount, quality and consistency) and extent of agreement within the literature that is reviewed. The summary qualifying terms the IPCC experts assign to the evidence evaluation are: *limited, medium or robust.* For level of agreement in the evidence, they are *low, medium, or high.* A synthesis of evidence evaluation and agreement then results in a level of confidence which is expressed using five qualifiers: *very low, low, medium, high and very high.*

Likelihood of occurrence

Likelihood is used to express a probabilistic estimate of the occurrence of a single event or of an outcome. It may be based on statistical or modelling analyses, elicitation of expert views, or other quantitative analyses.

The following terms have been used to indicate the assessed likelihood: *virtually certain* 99–100 % probability, *very likely* 90–100 %, *likely* 66–100 %, *about as likely as not* 33–66 %, *unlikely* 0–33 %, *very unlikely* 0–10 %, *exceptionally unlikely* 0–1 %. Additional terms (*extremely likely*: 95–100 %, *more likely than not* >50–100 %, *more unlikely than likely* 0–<50 % and *extremely unlikely* 0–5 %) may also be used when appropriate.

Source: Mastrandrea et al. (2010).

In Sects. 6.3 and 6.4 we examine in more detail adaptation and mitigation as they appeared in the 4th (2007) and 5th (2014) Assessment Reports.

6.3 Adaptation and Mitigation in the 4th Assessment Report (2007)

The IPCC published its 4th assessment report almost 20 years after the Panel was first established. During those intervening years, not only had the scientific basis of global warming as the driver of climate change progressed. So too had the thinking about how to respond to the challenge and we have seen above how the responses have coalesced under two broad headings: mitigation and adaptation. The 2007 WGI report was titled 'The physical science basis'; WGII: 'Impacts, Adaptation and Vulnerability'; and WGIII: 'Mitigation of Climate Change'.

6.3.1 Adaptation as Covered by the 2007 Report of WGII

WGIII defined adaptation as:

> The adjustment in natural or human systems to a new or changing environment. Adaptation to climate change refers to adjustment in natural or human systems in response to actual or expected climatic stimuli or their effects, which moderates harm or exploits beneficial opportunities. Various types of adaptation can be distinguished, including anticipatory and reactive adaptation, private and public adaptation, and autonomous and planned adaptation.

The above 2007 definition of adaptation related clearly to climate change as defined by physical science, which for the IPCC (WGI) was:

> Climate change refers to a change in the state of the climate that can be identified (e.g. by using statistical tests) by changes in the mean and/or the variability of its properties, and that persists for an extended period, typically decades or longer (Hegerl et al. 2007: 667).

Thus, in the IPCC's eyes of 2007, adaptation was intended to address the physical impacts of climate change, such as those resulting from changing weather patterns (for example, floods and droughts) and sea level rise. It is interesting that the adaptation definition also distinguished between adaptation of 'natural systems' and 'adaptation of human systems'. The report was not clear about natural systems. Did it mean changes in biodiversity of plant and animal life; did it mean migration to new habitats; or what? Figure 6.1 provides one example of what we assume the IPCC to mean by adaptation in 'natural systems'.

In any case, the emphasis in the report was on human adaptation which 'will be necessary to address impacts resulting from the warming which is already unavoidable due to past emissions'. On this, the 2007 report made some general points, for example (we provide our interpretation and comment—with the benefit of hindsight—in square brackets after each):

Fig. 6.1 Pink Campion (*Silene dioica*) in flower 19 December 2014 after an unseasonally mild autumn and early winter in the United Kingdom. This plant grows wild in several locations, including by the side of the road in many rural areas, as photographed here. It is found throughout northern and central Europe. Its normal flowering season is May to October

- Adaptive capacity is intimately connected to social and economic development but is unevenly distributed across and within societies. [*Our interpretation and comment*: poorer countries and societal groups have less capacity to adapt to climate change than do more affluent countries and societal groups].
- A wide array of adaptation options is available, but more extensive adaptation than is currently occurring is required to reduce vulnerability to future climate change. There are barriers, limits and costs, and these are not fully understood. [*Our interpretation and comment:* inadequately understood barriers, limits and costs mean that adaptation at present is primarily a case of muddling through on a case-by-case basis].
- The array of potential adaptive responses available to human societies is very large, ranging from purely technological (e.g., sea defences), through behavioural (e.g., altered food and recreational choices), to managerial (e.g., altered farm practices) and to policy (e.g., planning regulations). [*Our interpretation and comment:* we know what we could do if the barriers, limits and costs that were highlighted in the previous bullet point did not exist.]
- Vulnerable regions face multiple stresses that affect their exposure and sensitivity as well as their capacity to adapt. These stresses arise from,

for example, current climate hazards, poverty and unequal access to resources, food insecurity, trends in economic globalisation, conflict and incidence of disease such as HIV/AIDS. [*Our interpretation and comment:* vulnerability has many dimensions and climate change adaptation might seek co-benefits with these other stresses.]

- The projected number of people affected [adversely by climate change] is considerably greater under the scenarios where development is characterised by relatively low per capita income and larger population growth, than under higher per capita income and lower population growth. This difference is largely explained, not by differences in changes of climate, but by differences in vulnerability. [*Our interpretation and comment*: 'low per capita income and larger population growth' is a popular depiction of low level of development and vice versa for 'higher per capita income and lower population group'. In other words, development is the best way of achieving adaptive capacity globally.]

6.3.2 Mitigation as Covered by the 2007 Report of WGIII

WGIII defined mitigation as:

> An anthropogenic intervention to reduce the sources or enhance the sinks of greenhouse gases. (IPCC 2007a, IPCC 2007b: 53-62).

This 2007 definition of mitigation even more unambiguously reflected the dominant standpoint of physical science in defining climate change. Intervention to *reduce the sources* was equated with reducing the sources of carbon dioxide and other greenhouse gas emissions. Overwhelmingly this referred to the burning of fossil fuels to release useful energy on which human economic and social life depends. Intervention to *enhance the sinks* of greenhouse gases meant mostly actions to conserve forests and increase the cover where the processes of photosynthesis would capture carbon dioxide for plant food (see Box 1.1 and associated text in Chap. 1). It also meant grand technological schemes to remove carbon dioxide from the atmosphere and store it (now known as CCS—Carbon Capture and Storage—see also the 2014 Report analysed below and Chap. 11).

The report claimed that a 'wide variety of policies and instruments' is available to incentivise mitigation action. One such was, and remains a favourite option among western governments, to put a price on carbon emissions which would incentivise all economic sectors to innovate into 'clean' technologies. It also suggested that co-benefits could be achieved with respiratory health by reducing greenhouse gas concentrations in the atmosphere, and that changes in lifestyle, behaviour patterns and management practices can contribute to climate change mitigation across all sectors. (Ibid.)

6.4 Adaptation and Mitigation in the 5th Assessment Report (2014)

Fast-forward now to 2014 as this book is being written, where mitigation and adaptation continue to be entrenched within two of the three IPCC working groups (WGII and WGIII) as the two general approaches to responding to the challenge of climate change.

6.4.1 The 2014 WGII Report 'Impacts, Adaptation and Vulnerability'

Carrying the same title as in 2000 and 2007, WGII describes its brief as follows:

> The WGII contribution considers the vulnerability and exposure of human and natural systems, the observed impacts and future risks of climate change, and the potential for and limits to adaptation. The chapters of the report assess risks and opportunities for societies, economies, and ecosystems around the world. (IPCC 2014a)

The report simplifies its definition of adaptation from 2007 to emphasise human systems and actions:

> The process of adjustment to actual or expected climate and its effects. In human systems, adaptation seeks to moderate or avoid harm or exploit beneficial opportunities. In some natural systems, human intervention may facilitate adjustment to expected climate and its effects.

The above definition with its emphasis on the human reflects the stronger voice from the social sciences in the 2014 report compared to previous reports. It also introduces two new concepts to its analysis, each further reflecting a robust social science input:

- Risks (and risk analysis) where the report frames adaptation as a response to the individual and societal risks associated with climate change.
- Resilience, which it defines as, 'The capacity of social, economic, and environmental systems to cope with a hazardous event or trend or disturbance, responding or reorganizing in ways that maintain their essential function, identity, and structure, while also maintaining the capacity for adaptation, learning, and transformation'. This somewhat tortuous definition contains two basic ideas about resilience: (i) the capacity to adapt in order to cope with particular events, trends or disturbance at any given point; (ii) the capacity to evolve over time through learning.

We note that O'Brien and O'Keefe (2014) caution against uses of resilience which decontextualise away from socio-economic circumstances of poverty, inequality and vulnerability and call for people-centred, bottom up approaches, linked by social learning. The 2014 WGII report is not, however, unaware of these issues. It re-emphasises the 2007 concern with the ways in which climate-related hazards exacerbate other stressors such as poverty, stating: 'Climate-related hazards affect poor people's lives directly through impacts on livelihoods, reductions in crop yields, or destruction of homes and indirectly through, for example, increased food prices and food insecurity.' While noting the predominance of 'engineered and technological' options for adaptation, 'often integrated within existing programs such as disaster risk management and water management', the report recognises 'the value of social, institutional, and ecosystem-based measures and of the extent of constraints to adaptation'.

It goes on to claim that 'Governments at various levels are starting to develop adaptation plans and policies and to integrate climate-change considerations into broader development plans', and offers five climate-related reasons for concern or key risks (IPCC 2014a: 12) which we present in edited form below. All are stated within the range 'medium confidence' to 'high confidence' (Box 6.1):

1. **Unique and threatened systems**: especially ecosystems and cultures that are already at risk from climate change.
2. **Extreme weather events**: such as heat waves, extreme precipitation and coastal flooding.
3. **Distribution of impacts**: Risks are unevenly distributed and are generally greater for disadvantaged people and communities in countries at all levels of development.
4. **Global aggregate impacts** on, for example, the Earth's biodiversity and the overall global economy.
5. **Large-scale singular events**: the risk of abrupt and irreversible changes, for example, warm-water coral reef and Arctic ecosystems are already experiencing irreversible regime shifts.

The report then ends with a set of principles for effective adaptation, of which the first five are quoted directly below. All are expressed with 'high confidence' except the last which is summarised as 'robust evidence, high agreement' (Box 6.1; Ibid: 25, 26):

- **Adaptation is place- and context-specific, with no single approach for reducing risks appropriate across all settings.**
- **Adaptation planning and implementation can be enhanced through complementary actions across levels, from individuals to governments.** National governments can coordinate adaptation efforts of local and subnational governments, for example by protecting vulnerable groups, by supporting economic diversification, and by providing information, policy and legal frameworks, and financial support. Local government and the private sector are increasingly recognized as critical to progress in adaptation, given their roles in scaling up adaptation of communities, households, and civil society.

- **A first step towards adaptation to future climate change is reducing vulnerability and exposure to present climate variability. Strategies include actions with co-benefits for other objectives.** Available strategies and actions can increase resilience across a range of possible future climates while helping to improve human health, livelihoods, social and economic well-being, and environmental quality... Integration of adaptation into planning and decision making can promote synergies with development and disaster risk reduction.
- **Adaptation planning and implementation at all levels of governance are contingent on societal values, objectives, and risk perceptions. Recognition of diverse interests, circumstances, social-cultural contexts, and expectations can benefit decision-making processes.** Indigenous, local, and traditional knowledge systems and practices, including indigenous peoples' holistic view of community and environment, are a major resource for adapting to climate change, but these have not been used consistently in existing adaptation efforts. Integrating such forms of knowledge with existing practices increases the effectiveness of adaptation.
- **Decision support is most effective when it is sensitive to context and the diversity of decision types, decision processes, and constituencies** Organizations bridging science and decision making, including climate services, play an important role in the communication, transfer, and development of climate-related knowledge, including translation, engagement, and knowledge exchange.

The last two principles that we have quoted are of particular interest to this chapter because they concern knowledge—(i) indigenous, local and traditional knowledge systems, and (ii) the process of communicating, transferring and developing climate-related knowledge, which we assume to mean natural and social scientific climate-related knowledge. These two principles also underscore the 2014 report's concluding words: 'Underestimating the complexity of adaptation as a social process can create unrealistic expectations about achieving intended adaptation outcomes'.

6.4.2 The 2014 WGIII Report 'Mitigation of Climate Change'

This is the same title as 2007, but not 2003 when it was simply called 'Mitigation'. We speculate how much agonising over terminology must occur at IPCC meetings. It is very unlikely that the extra phrase—*of climate change*—was added without a reason.

The phrase was almost certainly added to clarify that WGIII's remit is specifically concerned with mitigation of the physical phenomenon that we call anthropogenic climate change. This might seem like a trivial clarification but it's important here to avoid confusion with other possible uses of the label 'mitigation' and to draw a clear dividing line between mitigation and adaptation. For example, much adaptation is intended to reduce the vulnerability of people to climate-related impacts. One could just as easily substitute the word 'mitigate' for 'reduce'. Both words mean 'lessen'. In other words adaptation may involve mitigation, but not mitigation in the sense that IPCC uses the word.

One has to be clear, therefore, about what one is mitigating and WGIII's current description does just that. Mitigation aims to address the physical drivers of climate change—global warming that results from increased greenhouse gas concentrations in the atmosphere. As already stated, adaptation aims to address the impacts (especially the human impacts) of climate change, impacts that are currently being felt and their exacerbation into the future.

This clear distinction between mitigation and adaptation has led some climate change campaigners to suggest that the two are in some kind of opposition, and that any emphasis on the latter is simply 'sticking plaster' to avoid more fundamental repair associated with the former. Most would concur, however, with the IPCC view that both adaptation and mitigation are necessary and that they are complementary rather than oppositional.

What then, does the 2014 WGIII report have to say about mitigation of climate change? First, its definition is a one-word variation ('human' for 'anthropogenic') from that used in 2007:

> Mitigation is a human intervention to reduce the sources or enhance the sinks of greenhouse gases. (IPCC 2014b: 4)

The following are some key findings of the WGIII report that are of interest to this book, together with our interpretation and comment on each. They do not, we stress, constitute a comprehensive summary of the group's report. Six of these findings are stated with 'high confidence' and one with 'medium confidence'. The remaining eight do not state confidence levels although four indicate the quality of the evidence and degree of agreement (Box 6.1). Numbers in brackets refer to the relevant page numbers of IPCC (2014b):

- Climate change has the characteristics of a collective action problem at the global scale, because most greenhouse gases (GHGs) accumulate over time and mix globally, and emissions by any agent (e.g., individual, community, company, country) affect other agents. International cooperation is therefore required to effectively mitigate GHG emissions and address other climate change issues (5). [*Our interpretation and comment:* greenhouse gases in the atmosphere do not know the meaning of political or any other human borders. Therefore, we have to act together.]
- Issues of equity, justice, and fairness arise with respect to mitigation and adaptation. Countries' past and future contributions to the accumulation of GHGs in the atmosphere are different, and countries also face varying

challenges and circumstances, and have different capacities to address mitigation and adaptation. The evidence suggests that outcomes seen as equitable can lead to more effective cooperation (5). [*Our interpretation and comment*: this is a re-statement of a justifiable complaint from developing countries that they are not historically responsible for the current level of greenhouse emissions yet they often bear the negative consequences and are expected to share the costs of mitigation.]

- Among other methods, economic evaluation is commonly used to inform climate policy design (5). [*Our interpretation and comment:* 'It all boils down to the economy, stupid!'. This comment is often heard during national election campaigns in the UK. Economic evaluation will follow the well-worn approach of assigning values to the costs and benefits of proposed mitigation options].
- Climate policy intersects with other societal goals creating the possibility of co-benefits or adverse side effects. (5). [*Our interpretation and comment:* co-benefits might include health, new 'green' jobs, etc. Adverse side effects might arise because the money spent on mitigation could arguably be put to better use directly on improving health, education, etc.]
- Without additional efforts to reduce GHG emissions beyond those in place today, emissions growth is expected to persist driven by growth in global population and economic activities. (9). [*Our interpretation and comment:* this is probably true with respect to growth in economic activities. As a corollary to our comment above on the subject of issues of equity, justice and fairness, low- and medium-income countries insist on their 'right to economic development' so that they may one day enjoy lifestyles comparable to those in high-income countries. Population growth is more difficult to assess. There used to be a slogan in Bucharest that read, 'Development is the best contraceptive' (Hewitt and Smyth 1992: 78). So, while it's obvious that more people on the planet will want to consume more energy, development presents the best long-term option for reducing the rate of population growth. The rich countries already have more-or-less stable populations.]
- Mitigation requires at least stabilising greenhouse gas emissions. (10ff). [*No further interpretation needed.*]
- Temporary overshoots in atmospheric concentrations of greenhouse gas emissions are likely, even with mitigation. They can be addressed through Carbon Dioxide Removal (CDR) technologies and methods. These include:
  - Carbon Capture and Storage (CCS)
  - Bioenergy crops which theoretically absorb as much carbon dioxide as they emit when burnt and are therefore classified as 'carbon neutral'
  - Afforestation (Box 1.1 in Chap. 1).

Risks are associated with all of these. (13). [*Our interpretation and comment:* CCS already exists but is far from widespread and hence unproven; bioenergy crops have been criticised for taking up land that could be used for food crops (see below); afforestation could lead to displacement of indigenous groups and their livelihoods.]

- Estimates of the aggregate economic costs of mitigation vary widely and are highly sensitive to model design and assumptions as well as the specification of scenarios, including the characterization of technologies and the timing of mitigation (15). [*Our interpretation and comment*: yes, economics is not an exact science. Despite numbers being assigned to costs and benefits, they depend on assumptions rather than value-free calculations. See Chap. 3 and our discussion of the debate that followed the Stern Review 'The economics of climate change' that was produced for the UK Government, where Stern's assumptions about discount rate were challenged.]
- Mitigation scenarios that limit greenhouse gas concentrations in the atmosphere to about 450 or 500 ppm CO_2eq[1] by 2100 show reduced costs for achieving air quality and energy security objectives, with significant co-benefits for human health, ecosystem impacts, and sufficiency of resources and resilience of the energy system (17). [*Our interpretation and comment:* 450–500 ppm CO_2eq represents the most feasibly achievable concentrations that won't cause catastrophic disruption (Stern 2010: 39–41). Overall, this point from the WGIII Report adds detail to the co-benefits point above. 'Energy security objectives' and 'resilience of the energy system' refer to the possibility of countries becoming self-sufficient in their energy needs through developing renewable energy alternatives and no longer needing, for example, to import oil. Thus renewable energy development and national energy security become a win-win and another co-benefit.]
- Efficiency enhancements and behavioural changes, in order to reduce energy demand compared to baseline scenarios without compromising development, are a key mitigation strategy in scenarios reaching atmospheric CO_2eq concentrations of about 450 or 500 ppm by 2100. (21). [*Our interpretation and comment:* efficiency enhancements and behavioural changes are more easily said than done. The report suffers here from lack of in-depth social science analysis on the feasibility and desirability for different societal groups.]
- Behaviour, lifestyle and culture have a considerable influence on energy use and associated emissions, with high mitigation potential in some sectors, in particular when complementing technological and structural change. (21). [*Our interpretation and comment*: again, the implied

[1] CO_2eq stands for carbon dioxide equivalent. It represents the combined concentrations of all greenhouse gases in the atmosphere, measured in parts per million (ppm), adjusted for their global warming effect in relation to carbon dioxide.

behavioural, lifestyle and cultural changes are more easily said than done. Our lived experience perspective in Sect. 6.5 will highlight the issues.]

- Regional differences influence the choice of transport mitigation options. Institutional, legal, financial and cultural barriers constrain low-carbon technology uptake and behavioural change. Established infrastructure may limit the options for modal shift and lead to a greater reliance on advanced vehicle technologies; a slowing of growth in light-duty vehicle demand is already evident in some OECD countries. For all economies, especially those with high rates of urban growth, investment in public transport systems and low-carbon infrastructure can avoid lock-into carbon intensive modes. Prioritizing infrastructure for pedestrians and integrating non-motorized and transit services can create economic and social co-benefits in all regions (23). [*Our interpretation and comment:* in high-income economies, radical shifts towards lessening dependence on private motor cars depend on behavioural and possibly lifestyle changes which are difficult to achieve. This is despite public transport systems functioning in these economies. High urban population growth in developing economies seems to swamp attempts to create public transport infrastructure and institutional capacity is often generally lacking.]
- Lifestyle, culture and behaviour significantly influence energy consumption in buildings... For developed countries, scenarios indicate that lifestyle and behavioural changes could reduce energy demand by up to 20 % in the short term and by up to 50 % of present levels by mid-century. In developing countries, integrating elements of traditional lifestyles into building practices and architecture could facilitate the provision of high levels of energy services with much lower energy inputs than baseline. (24). [*Our interpretation and comment:* see above concerning the challenges of lifestyle and behavioural changes in developed/rich/high-income countries. Meanwhile, the possibilities offered for developing countries are no more than aspirational, with no analysis of how they might be achieved.]
- Bioenergy can play a critical role for mitigation, but there are issues to consider, such as the sustainability of practices and the efficiency of bioenergy systems. Barriers to large-scale deployment of bioenergy include concerns about GHG emissions from land, food security, water resources, biodiversity conservation and livelihoods. The scientific debate about the overall climate impact related to land-use competition effects of specific bioenergy pathways remains unresolved. (26) [*Our interpretation and comment:* see our previous comment on bioenergy under the 'Temporary overshoots' bullet point. To illustrate, Toulmin (2009: 134) reports how thousands of small-scale rice farmers face eviction in the Wami of Tanzania which has been earmarked for sugar cane as a bioenergy crop].
- The largest mitigation opportunities with respect to human settlements are in rapidly urbanizing areas where urban form and infrastructure are not locked in, but where there are often limited governance, technical,

financial, and institutional capacities. The bulk of urban growth is expected in small- to medium-size cities in developing countries. The feasibility of spatial planning instruments for climate change mitigation is highly dependent on a city's financial and governance capability. (27) [*Our interpretation and comment:* Yes, this reinforces our comment about transport systems in rapidly urbanising areas above. The governance, technical, financial and institutional capacities represent formidable challenges. Toulmin (2009: 108), for example, warns of the danger and costs of assuming that actions can be grafted onto a network of existing infrastructure which does not exist in many African countries.]

If we compare the above with the highlights we drew on mitigation from the 2007 WGIII report, we find, interestingly, that carbon pricing, of which much is made in the 2007 report, is referred to in the context of various mitigation strategies in the 2014 report, but is not addressed as a subject in its own right. We can only conclude that the 2014 report assumes that the debate is closed, and carbon pricing as a driver of innovation to mitigate climate change is assumed throughout.

Beyond that, we see in the 2014 and 2007 reports two of the same themes: establish co-benefits, and promote lifestyle and behavioural changes. They echo important themes raised by Hulme (2009, 2013) who stresses both co-benefits and re-thinking what we value as human beings. We return to these in Sect. 6.5 alongside additional highlights from the 2014 report, namely: the need for collective action, differential mitigation that is sensitive to the development needs of developing countries, institutional hurdles, challenges and economic analysis of costs and benefits. While most of these feature in one form or another in the 2007 report, they are less-developed and/or implied rather than explicit.

6.5 The Lived Experience of Mitigation and Adaptation

The behavioural and lifestyle challenges for effective mitigation that are highlighted in both the 2007 and 2014 IPCC reports are the reverse side of the coin to the legitimacy issue of climate change action that we raised in Chap. 5. Populations will not easily be convinced to alter behaviours and lifestyles that will inevitably take them out of their comfort zones. The chances are, therefore, that attempts by governments to persuade their citizens will be perceived as not legitimate and against the wishes of the people. In these circumstances, 'the people' will in all probability oppose such attempts through the usual democratic channels where they exist, or even outside of them.

This is more an issue for mitigation interventions than it is for adaptation, or at least for adaptation that reacts to an event. If one is victim of an extreme weather event, as we recorded in Chap. 1, then one is unlikely to oppose adaptation measures to improve flood defences in the area. Even if one is not directly involved, the media images of displacement and suffering will usually be enough to persuade at least not to oppose remedial action. Sometimes the mere threat of a disaster that could arrive at any time, including the present, is enough to legitimise. The building of the River Thames barrier to protect London from flooding is a good example. Thus, with reactive adaptation to events or even pro-actively to threats of events, we can muddle through, albeit with immense costs to some people for whom the adaptation comes too late. Yes, in all probability it will constitute a piecemeal response, but it is less likely to lock us into a pathway that could have unintended consequences lurking at every corner.

Nor should we lose sight of the fact that, in poor developing countries, people are never passive and are adapting anyway. Harmer and Rahman (2014), for example, review how farmers in these countries are forced continually to adapt within whatever constraints pertain. In doing so they may locate their responses within broader livelihood strategies, including ‘normal climate variations’, rather than as adaptation to climate change per se. These authors also note ‘a range of strategies from financial responses, to agricultural changes, to religious and cultural strategies, to the use of local and wider support networks’, and the influence of their socio-economic contexts (Ibid.). As we argue in Chap. 5, a lived experience lens draws attention to all of these dimensions by illustrating what is happening on the ground in places and their contexts.

Mitigation of climate change, however, is difficult to pin down. Many impacts are projected for the future, and many of today’s adults perhaps won’t be dramatically affected by them. In other words, the impacts are unlikely to be experienced by most of the current generation and we might personally escape them. Thus we feel that the future can look after itself. Moreover, even when an extreme weather or tidal surge event does occur in the present, citizen suggestions that it is linked to climate change are met with a scientific chorus that one cannot attribute any single extreme event to this phenomenon, despite climate change models indicating their increasing likelihood (see Chap. 3). With all these doubts concerning future uncertainty and present difficulties of attribution on one side, on the other there are common-sense indications that a proposed intervention to mitigate climate change may affect us negatively in our current lives in one way or another. As we reflect on these matters, there can be only one winner.

This might seem substantially a problem of affluence, because when it comes down to it we associate mitigation with some form of detrimental lifestyle change (minor or major). We don’t want to change affluent lifestyles, and especially not for something that feels so vague, something we cannot hold, but only feel and anticipate the possible effects. Mitigation is also an issue for poorer societies, however, for they understandably and rightly aspire to the level of affluence that they see elsewhere and which is in their faces through advertising hoardings and other media every day of their lives. This is why, at a national and even regional level, poorer countries demand that their right to development is not compromised. They will not legitimise international mitigation actions that do so compromise, unless of course fully compensated.

The two sides of the coin—the challenge of lifestyle/behavioural change and the challenge of legitimising mitigation actions—relate to our lived experiences of climate change. They concern our reflections on what we value (current lifestyle or a lifestyle we would like to have), our common sense reflections on what works in our favour and what might not, and our thought processes as conditioned by our socio-economic circumstances. This is why it is difficult to engage positively with mitigation. Our lived experiences tell us of likely loss and hence sacrifice: loss of amenity, loss of employment for many as the economy adjusts and jobs in traditional energy and other sectors are replaced, loss of forest livelihoods and displacement of groups that live there, displacement of farmers from land that is now earmarked for bioenergy crops, and so on. These are just the tip of the iceberg of sacrifices many may have to make, all in the cause of an uncertain future which, even if it comes to pass, might not affect us directly.

Of the case studies that we have presented in this book so far, two stand out as examples of mitigation of climate change, or attempts at mitigation. They appeared in Chap. 1 where both were linked to harmful impacts in the present:

1. The exacerbation of vulnerability among two East African forest dweller communities by conservation actions to create carbon 'sinks' (to use the terminology of the current IPCC definition of mitigation—see Sect. 6.4.2);
2. The so-far successful attempt by a community in England to resist the creation of a wind energy farm in the vicinity, which would, among other effects, give rise to loss of visual amenity, create noise disturbance, lower house prices, and be harmful to local nature.

Despite being very different in terms of the physical mitigation action and socio-economic context, and being separated by great distance, these two case studies have one important thing in common: mitigation of climate change has been a negative experience or is anticipated to be a negative experience.

Let us now explore these issues of required behavioural changes/personal lifestyle threat and legitimacy of mitigation action through two further examples.

Example 1: What can I do personally to save the planet from climate change? Lived experiences of concerned citizens in Europe

There are many concerned citizens who have never experienced a drought or a flood or the encroachment of the sea, but who nevertheless aspire towards climate-friendly lifestyles. They are concerned, less for themselves, but for the world in which their children will grow up, for future generations generally who will bear the brunt of our carbon-profligate actions today, and for those who already are experiencing the impacts of climate change, especially those who are poor and are less able to adapt. While government incentives to change may play a part in the behaviour of concerned citizens, much of it arises from their values, values which relate to fairness and justice for both future generations and poor people in the present (see Chap. 11 for further discussion of values).

Between 2009 and 2012, the authors of this book were part of a pan-European team that created three e-learning modules at Masters degree level on the lived

Table 6.1 Carbon footprint per capita of selected countries (corresponding to countries of students who participated in the exercise), industrialised countries and worldwide

	Average per capita carbon footprint (tonnes of CO_2 emissions per annum)
Cameroon	0.41
France	6.2
Germany	9.8
Indonesia	1.69
Netherlands	8.7
Portugal	5.63
Spain	7.7
United Kingdom	9.8
Industrial nations	11
Worldwide average	4
Worldwide target to combat climate change	2

Source Carbon Footprint Ltd. (2015)

experience of climate change (See Box 6.2 for a fuller description of this project). The modules were piloted to 26 students from the participating universities in 2011. As part of the pilot we set a task for the students to measure and compare their individual carbon footprints, that is, the number of tonnes of carbon (mostly as a chemical constituent of carbon dioxide) that they generate personally per annum. They were directed to a free 'carbon calculator' on the internet for this purpose. The site provided for comparison purposes the average per capita carbon footprints of the country in which the user lives, the average for affluent, industrialised nations, the average worldwide footprint, and a target figure of 2 tonnes of carbon per annum which represents the global average if we are to hold global warming to no more than 2 °C above pre-industrial levels (Table 6.1). The calculator asked users questions concerning home energy use, domestic and international travel by different means, and lifestyle in terms of food and other purchases.

Box 6.2 The European Union Erasmus Project: The Lived Experience of Climate Change: e-Learning and Virtual Mobility (LECHe)

This project created postgraduate curriculum and virtual learning communities in relation to climate change between 2009 and 2012. It aimed to:

- Contribute to an informed, active European citizenry and the United Nations Decade on Education for Sustainable Development
- Inform European Union policy.

The project focused on the lived experiences of climate change, conceived in this early project as how individuals, communities and organisations conceive and respond to the perceived local impacts (for example, rising sea levels, extreme weather, biodiversity changes) of the phenomenon.

It involved collaboration between eight universities plus the umbrella European Association of Distance Teaching Universities, and was funded by the European Union Erasmus Programme. It spanned six European Union member countries.

The educational level of the project was postgraduate Masters. A key feature was that the curriculum complemented existing and proposed Masters programmes in the area, rather than attempted to create a parallel programme. After development as distance teaching resources and piloting to 26 students drawn from the participating universities, the modules became open educational resources at the project end in April 2012.

Any university anywhere in the world is free to use or adapt the modules within their own programmes, through their normal accreditation processes. They may be used flexibly by students: as available educational resources without assessment or accreditation to enhance their studies, or as conventional modules with assessment and accreditation. The modules may be accessed at: http://www.open.edu/openlearnworks/course/view.php?id=1767.

The activity provoked a lively discussion on the student discussion forum. This is how it went during its final days:

Mariana (German)

To end my thoughts: We can do a lot individually (move to smaller houses and insulate them, install photo-voltaic-panels, be very prudent consumers, use public transportation if available, etc.) and by making a deep analysis of one's own way of life and to think about, if a lot of our behaviour is really necessary. I have to question myself. For example, is it really necessary that my kids go at least twice a week to sports and once a week to a music course, what makes at least six rides. I doubt that it is really necessary for their development, but as all mothers I know offer this to their children, I fear, that my kids could have a disadvantage. Another question is: Is it really necessary to live in a house with 200 m^2 or would a smaller flat, but then in the city with better access to public transportation, also be sufficient? But I think a lot has to be done by changing our systems and structures, that means for my personal situation, the system we live in has either to provide an efficient public transportation even in rural areas or has to provide some housing acceptable for families. As this list could be endless, I would like to stop here.

Star (British)

It is fascinating to see the differing lifestyles we all lead. What interests me are the commonalities across such difference.

First, I was impressed how many people are already making significant changes to their lifestyle towards carbon efficiency. However, despite this many of us are close to the national average [for European countries] which is surprising.

Most of us face different but equally understandable structural reasons why certain components of the footprint cannot be reduced. Whether it's needing to fly to visit family, or choosing to live in a rural area making the car a necessity or even taking a long haul flight to deliver a conference paper, all of these things seem reasonable choices in the current perspective. Do we need to give up such 'liberties' in order to reach the global target?

I see the issue here to be a conflict once again between agency and structure, how much can we change through our own agency (choices) and how much does it rely on structural changes? For example, people could still do the things above if air travel and car travel were possible with significantly reduced emissions.

Second, I've been interested to see the variety of emotions expressed in these responses such as embarrassment and shock. Perhaps this is because we are sharing intimate details of our lifestyles with each other in a group where low emissions are valued more highly.

Luis (Portuguese, living on an off-shore island, one of the Azores)

My footprint is 9.79 t., much larger than the average in Portugal (5,63 tonnes). I think it is a consequence of living in an island in the middle of the Atlantic Ocean. We have the need to use the plane between the nine islands (we can use the boat only in summer) and the mainland. During last year I travelled twice to Lisbon to work.

In São Miguel Island (where I live) there's no rail and the bus system is not very good and because I live in a rural location, I'm obliged to travel by car.

Here it is not so easy to find local products, because they are very few (fish, meat, dairy products and some fruits and vegetables). Everything else comes from the mainland. At home we make waste separation, but all the waste goes to mainland to recycle (A huge impact if contemplated, I think!).

The house footprint is not so big because the Azores have a temperate weather, with average temperatures during all the year.

Sandra (South African, living and working in Germany)

I have compared my emissions to that of Mariana, since she also lives in Germany. Despite this fact, we lead very different lives, and have different emissions (hers are much lower than mine!). Perhaps I can find a happy medium between her activities that reduce her emissions, and my own, so that I can bring down my amount.

First, Mariana has children, and I do not. It also means that she requires a car, and needs to drive around more. I do not even own a car, so I cannot compromise there.

I found it interesting that she identified living in a village as a reason for more emissions, and it does seem like that is the case. I, on the other hand, live in the middle of Cologne city, and have a lot of infrastructure to assist me in lowering my transport emissions, which I do make good use of. It does seem wrong though, that a more green infrastructure cannot be put in place in small towns- at least in terms of the biking lanes. I believe this is something the government should look into. Coming from Johannesburg, I understand how frustrating it is when you have no choice but to drive everywhere, no matter how good your intentions.

I then looked at my highest emissions, flights, which I need to reduce. This is very hard, unfortunately, as my work does not currently have the capacity to reduce my amount or work-related flights. Perhaps I can lobby, and get other colleagues to join me, in ensuring that more meetings are done electronically, or over telecom. Another way I could reduce my emissions in this regard is by moving back to South Africa, so that I do not fly back and forth to see my family there.

Finally, I could try to reduce the household emissions. I have a really small flat, unlike Mariana who probably has a bigger house due to having a family. I could, like her, try to think of extra ways to reduce energy consumption in the home, particularly in summer with the long light days.

Regina (German)

When I imagine that our results are the ones from people who do already care about reducing their impact on our environment, I become frightened how the others are looking like.

And as I know that even if people understand things with their brain, they are still far away from changing their habits. This I see always on my own behaviour.

Kerry (British)

In comparing my results with the average for [poorer] countries such as Cameroon and Indonesia I was shocked and ashamed at the unsustainable and energy intensive lifestyles we adopt in the UK. So much of our carbon consumption appears to often come from our overall lifestyle and culture. This may seem like an obvious point but after reading everyone's separate posts it appears to me that the many of us have very similar final carbon footprints even though the individual breakdowns may be very different—for example, some of us may fly a lot, or use the car or have high housing costs. Therefore, it occurred to me from this that the devil is not so much in the detail but in our overall behaviour. This raises a fundamental question when it comes to lowering carbon footprints in 'industrialised' countries that what needs to change is not necessarily how often we use the car or whether we buy local, organic food but that our overall attitudes and way of life need to change in order to become more sustainable.

I found this to be particularly interesting when reading Olivier's [originally from Cameroon, now living and working in Germany] post who—because he has moved to a more carbon intensive culture—his footprint has increased. However he has also maintained the energy efficient behaviours that have helped keep his footprint significantly lower than the national average. This is also evident in Phillip's [German, now living and working in Indonesia] interesting post where his behaviour has resulted in a higher than average carbon footprint because of his ability to travel and also his secondary behaviour being more carbon intensive compared with local Indonesians. Having always lived in a certain culture I think of myself as being environmentally aware but it appears that I am still lacking these fundamental behaviours and attitudes which would make me more carbon efficient. This is especially pertinent for me as my secondary emissions were so high and contributed over 50 % of my footprint. This suggests that my consumer behaviour is not as environmentally conscious as it should be in comparison with more sustainable cultures and I should make more of an effort to buy local rather than global. In tackling carbon efficiency on a global scale it would appear that these are lessons which carbon intensive nations also need to learn in order to significantly improve our overall average in the future.

Mirroring the point raised by Regina I find it concerning that these are the results of people who are already environmentally aware and who make a conscious effort to be energy efficient. This makes for quite a challenge if everyone in the world is to reach the target of 2 tonnes of CO_2 per year.

Phil (British)

I've found this forum to be fascinating and equally thought provoking. In reading the posts and considering the contributions, it is clear that I have a carbon footprint well above most here and the average for the UK.

I was interested to read Star's post as we both live in the UK and with two adults and two children living in the same property; consequently we have very similar footprints for energy consumption and also the secondary emissions are comparable as well. Maybe as Kerry points out, the results are a reflection of our general behaviour at a national level and not so much our individual patterns of energy use and consumption.

Whilst I'm certain that my consumption patterns could be improved, there are certain areas where I really struggle to see how I could significantly reduce my carbon footprint. This is most true when considering travel. The UK's public transport infrastructure has many limitations and I don't feel I could function effectively within my current work role if I had to rely on the train and bus networks, particularly living in the east of England. It is also pertinent that much of my travel is in pursuance of reducing carbon emissions within a large multinational organisation; an area where we have been relatively successful. Interestingly, this requires that all employees record their travel emissions, so the results here should really come as no surprise to me. I guess I have been looking at the 'bigger picture' which is a magnitude greater than my own personal emissions. No excuse I realise!

Katharina (German)

What fascinates me the most is that the students of this group differ so much in their lifestyles. We all live in totally different places, urban or rural, industrial or developing countries, in different phases of life and all kind of jobs.

Each lifestyle seems to come with specific difficulties regarding the attempt to live a climate friendly life. Some seem to be very contradictory, for example the fact that living in the countryside enlarges the carbon footprint as Mariana says. When comparing my results in detail with those of the others, it becomes clear to me that I really benefit from my current life situation in terms of a climate friendly life: I am relatively young and so far don't have kids or a huge house. I love the fast city life with its many possibilities and so far have no such reasons as the need to afford a huge apartment or a house for a whole family to move to the countryside. Ironically, with regard to the climate this works for my advantage.

I agree with Kerry that the reason for our overall relatively high carbon footprint is our high living standard: We have huge carbon footprints simply because we can. Those of us living in developing countries like Phillip can afford to go back home [Germany] to visit friends and family, those of us living in the countryside can afford to compensate the bad infrastructure with a car and I can go to different places in the world by plane if I feel like getting to know different cultures. I'm afraid that most of us are willing to cut back on some things, but not on that much as to reach the goal of 2 tonnes of CO_2. What do the others think about that? Do you agree with me or do you have a different opinion?

Kaori (Japanese, living and working in the Netherlands)

We live in affluence. We can make choices. It is worthwhile to continue thinking about better choices. But how far we are ready or able to give up our own "liberties" (according to Star) to achieve a drastic reduction in our emission?

Julie (British, living and working in Spain)

Our mains supply [in Spain] used to come directly from an underground spring, but the aquifer is so polluted by chemical residue from intensive agriculture that it's unsafe to drink and the municipal water company took it over four years ago. They've even imposed a new municipal 'residual waste' tax on all residents to help foot the clean-up costs. The tap water we get now is so heavily chlorinated and fluorinated that it even makes your food taste horrible so we have resorted to buying 25 litre bottles of water for drinking and cooking purposes. I am confident that this is not just an expatriate thing. The vast majority of my friends are Spanish and I don't know one single person who drinks the tap water here.

This shows how our lifestyle choices can be indirectly influenced by external factors (such as intensive food production) that we may vehemently disagree with and which, standing alone, we have little personal agency to do much about.

Star's snippet of lived experience as a parent made me think about how having children affects an adult's personal agency, particularly where transport choices are concerned—when you have children you have to take their needs and physical limitations into consideration as well as your own.

This led me on to think more about our own choice of family holiday this year—if we'd have flown to Switzerland and stayed in a hotel, the plane, trains, busses and hotels would have been operational anyway so although we'd have included this in our carbon footprint, we wouldn't have made any additional contribution to global emissions. In choosing to drive and camp we actually added about 0.8 metric tonnes of CO_2 to the global total. However, this choice was both a practical and economic one that enabled my husband to fulfil a lifetime dream of seeing the north face of the Eiger (he's a climber) for his 40th birthday, and enabled the whole family to spend some time with my brother who lives in the south of Germany (we usually holiday in Spain). If we hadn't driven, we wouldn't have been able to afford to go. Economic factors also, therefore, constrain our choices.

******

The above discussion brings into sharp focus a constant tension between what we feel it would be necessary to do to reduce drastically our personal carbon footprints, and why we find the task overwhelming, if not impossible. Where we live and what we live in, the transport infrastructure that is available, our responsibilities towards those who are close to us such as children and distant relatives, our jobs—all of these combine to limit radical changes to behaviours and ways of life, necessary as these changes are perceived to be. Relatively affluent people have a degree of choice as to the lifestyle we lead, but having made that choice we soon become ‘locked in’ to use another piece of terminology of the IPCC (see Sect. 6.4.2 above). Thus, we might choose to live in the countryside rather than the city where we work because it’s a better life for our children that we have also chosen to have. Because of that we require a reasonably large house and a car. We owe a lot to our children and have to take them many places by car and if we don’t they will feel deprived relative to their peers. We also have responsibilities to our parents who live halfway across the world from us, so we have no option but to fly to them.

In short, we do not have absolute free will, we are constrained by circumstance. The lived experience of the concerned, relatively affluent citizen is the experience of this tension.

The Nobel prize-winning economist Sen (1999) has written extensively on the freedom to choose to do what we value. What we value is of course highly subjective and diverse. In the above discussion, for many participants it might mean raising children in a happy, secure environment. On the other hand, for the Batwa and Ogiek communities of East Africa, and the subject of Nyukuri (2013) thesis, it might be to continue using the forests as a source of livelihood and well-being, and more generally to retain their identity. Put another way, it is the freedom we have to exert agency. Vulnerability, however, is the opposite of this. It is the lack of freedom to choose to do what we value, to exert our agency.

Sen’s main concern is development as freedom and the relative lack of it among the poor and vulnerable of the world. Nyukuri (Ibid.) in fact used his work to develop the framework for her own research into the climate-related vulnerability of the Batwa and Ogiek people. Sen defines his overall approach as the ‘capability approach’ where capability is the “real opportunity that we have to accomplish what we value” (Sen 1999: 74). This is essentially what he means as ‘substantive freedom’ as opposed to formal freedoms enshrined in the laws of the land in which one lives. Capability, therefore, concerns the means, tools and assets that enable a person to pursue what she/he would like to be or do (Nyukuri 2013: 77). In this sense, capability appears to be linked to a key concept that appears in the 2014 IPCC WGII report—resilience (Sect. 6.4.1 above).

The students who participated in the LECHe project carbon footprint exercise and discussion are either European or live and work in Europe. Compared with the Batwa and Ogiek of East Africa, nobody would ever think of analysing their lives and livelihoods in terms of vulnerability. Relatively speaking, they do possess the capability to exert their agency and accomplish what they value. They are resilient in the IPCC’s usage of the concept. Having made their choices, however, their freedom invokes responsibilities, thus immediately putting limits on that freedom. Sen makes this point, but it was also made many years ago by the French philosopher Albert Camus in his book ‘The Rebel’ (Camus 1953, 1962). To be free invokes a responsibility to respect the freedom of others. That places a limit to our freedom, a limit which we all must recognise to live within.

Substantively, for the students who participated in the above discussion, their responsibilities result in imperatives to do this and not that, to act in this way and not that way. Even if we are deeply concerned about climate change (as all the students who participated in the LECHe project were) and we would all value reducing significantly our personal carbon footprints, our lifestyle choices have limited our freedom to do so. Once locked into their logic and imperatives, changing our lifestyle and behaviours is very difficult, especially as doing so would likely occur at the expense of other things that we value. That is what our lived experience and our engagement with other lived experiences tells us.

Example 2: protesting again against wind farms
Wind farms are promoted in the UK (and elsewhere of course) as part of a renewable energy policy that has co-benefits. They are intended both to improve the UK's energy security (less reliance on imported oil and gas) and to reduce its carbon footprint. In many places, however, they are opposed by local communities on amenity grounds, because of their visual intrusion on the landscape and the noise disturbance from the turbines for those living nearby. We saw this in the case study in Chap. 1, although also note that other quasi-scientific arguments were used in this example concerning the effectiveness of wind as a source of renewable energy. The politics of opposition will always draw on 'expert' evidence to support a case, if only to neutralise the expert evidence of proponents (see Chap. 3). Usually, however, the source of opposition is rooted in personal and communal experience and not science.

To illustrate, here is another example of opposition to a wind farm, taken from a British regional newspaper (Box 6.3). In the article the community that will be directly affected if the wind farm goes ahead has only recently found out about the plans. There has, therefore, been no time as yet to research and marshal expert counter-evidence to the proposal. The voice we hear in this article, therefore, comes from raw concern about what the local residents value becoming devalued. The following quotations exemplify:

> Our homes look out onto fields and pretty rolling hills countryside and now all we're going to see is these two big monstrosities [the turbines].
>
> I think it will have an affect on our property prices.
>
> The reason we moved up here was for a bit of peace and quiet, but now when we're sitting out in the garden we're going to hear the drone of wind turbines going round.

As with example 1, we hear in this example a voice of relative affluence.

Box 6.3 Extracts from a local newspaper report on opposition to a wind farm proposal in the UK

Villagers' anger at 150 ft wind turbine plan
By Emma Davidson, News Reporter.

Villagers have slammed plans to build two massive wind turbines near to their homes.

Plans for the development on farmland…include two 47.5 m (155 ft) tall turbines.

[A private sector company] wants to construct the… Wind Energy Development at the edge of the rural village.

But those living nearby have described the plans as "monstrous".

They are planning to oppose the plans by writing to Kirklees Council [the local municipal authority responsible for planning]…

[Resident] Katie Bottomley, whose family home is just 600 m away from the planned development, said: "… I know a lot of people will be objecting to this.

"Our homes look out onto fields and pretty rolling countryside and now all we're going to see is these two big monstrosities."

In plans submitted to the Council [the company] says that the development would save up to 469 tonnes of CO_2 emissions each year.

It says that the turbines are large but… "could be accommodated in the visual context of the wind energy development.

"This is largely due to the scale and character of the receiving environment and other built influences of a vertical nature—including the nearby Emley Moor Mast [a 330 m television tower a few kilometres away]."

But Katie, who works for an environmental consultancy firm… said: "I'm all for renewable energy but not a cost to the village.

"These turbines are massive and would be on top of a hill… so they will be… [the only thing] many of us… will be able to see when we look out the window.

"I think it will have an effect on property prices.

"I also don't think it's right that it should be approved just because of Emley Moor Mast, because that's been there forever and is hardly the same thing."[2]

Katie is also concerned about noise and potential safety risks…

She said: "The reason we moved up here was for a bit of peace and quiet, but now when we're sitting in the garden we're going to hear the drone of wind turbines going round.

"Also when it snows, ice can build up on the propellers and who knows what will happen if that starts flying off…"

[Source: Huddersfield Daily Examiner: 5 July 2010]

Note also in this article that the company that has applied for planning permission to build the turbines provides a technical reason to justify their location close to this village. It is that a high mast is already in the vicinity. This is a local-level example that mirrors the technical and largely apolitical arguments of the IPCC concerning mitigation and adaptation.

[2]Book authors' note: the original television tower on the site was erected in 1956.

Example 2 as a whole, however, illustrates that there is a politics to everything and in climate change it exists at the local as well as at national and international scales. It emerges significantly where mitigation interventions are going to affect the lived experience of a community of relatively affluent people who have made the choice of a particular location as their home. They are not going to accept the intervention and there is a record of successful local opposition to wind energy throughout the UK. In the final analysis, politicians at higher levels have to take note in the interests of governmentality (Chaps. 5 and 10). In other words, they have to seek legitimacy for their decisions that affect localities. Thus, adaptation and mitigation analysts need to realise that, while there are of course technical issues and disagreements, the major challenges concern the politics and legitimacy of interventions. Hearing and reflecting on lived experience can help them in this task because it brings the politics of decisions out into the open, stripping them of their technical cover.

Finally, both examples 1 and 2 appear to relate to debates about climate change being framed as a 'win–lose' game (Jaeger et al. 2012) where action is essential even if it means that some are harmed by it. Frames in this sense mean the mental (or cognitive) boundaries that determine how we analyse and generate knowledge, and intervene on a problem such as climate change. Within a particular frame are a set of key, inter-related concepts that act as analytical tools. For example, a classical economics frame would contain (among others) concepts of markets, competition and innovation which we may use to help analyse economic phenomena. Generalising, and putting the key point about frames slightly differently, they provide the cognitive structures by which we order information (Kempton 1991).

Jaeger et al. (2012) argue for a fundamental shift in climate change debates from a 'win–lose' to a 'win–win' frame, 'perhaps as a system transformative attractor for new green growth, sustainable development and technological innovation' (Ibid: Foreword). In this, they reflect economic analyses that have been marshalled in recent years to argue that changing to a low-carbon economy will bring long-term prosperity and new employment opportunities (see Chap. 3 and Stern 2010). From the current chapter, co-benefits with other pressing global societal challenges, as indicated in the 2007 and 2014 IPCC reports (Sects. 6.3 and 6.4 above) and Hulme (2009, 2013), are also examples of a win–win frame.

The analyses of Jaeger et al. and Stern, however, place their win-win scenarios within a deeper frame that has existed since climate change became a global societal issue in the late 1980s. This is that climate change is a deep threat and consequently a war (or at least the language of war) must be waged against it (a frame that Hulme explicitly reflects on). Thus, Stern's book refers to saving the world (sub-title) and that 'the risks are clearly enormous' (Ibid: 37). Jaeger et al. may claim to be setting a new win-win frame, but chapter authors in their volume clearly subscribe to the deeper threat and war frame. For example:

> It is clear that climate change is the greatest single threat to human development (Hobbs 2012).

> Climate change represents an unprecedented challenge (Hasselman and Voinov 2012).

> While the recognition of climate change as a major threat faced by humanity is beyond dispute… (Mangalagiu and Cameron 2012).

The problem with a 'threat and war' frame is that, like all wars, it demands sacrifices in the present. Insofar as the protagonists offer 'win–win' it is a vision of the future that makes the war worth fighting. Thus, in the climate change war it is a vision of a prosperous and sustainable future. Yet, climate activists are remarkably silent about the human processes of change and the sacrifices that might be required in the present in order to achieve the possibly golden future. 'Sacrifice' is a noble action, 'loss' is often ignoble, so the former substitutes the latter, but substantively the effect is the same. Loss/sacrifice applies to impoverished forest communities of East Africa as it does to affluent people living in England. It is writ large in the reflections of concerned citizens over Europe who, for better or worse, are locked into their lifestyle choices, and it is writ large in the shock felt by others as a renewable energy project threatens to disrupt their peace and quiet.

In his 2009 book, Mike Hulme, also draws critical attention to the deep frame that climate change is a battle or war when he concludes that 'Rather than placing ourselves in a "fight against climate change", we need a more constructive and imaginative engagement with the idea…' (Hulme 2009: 361). We agree wholeheartedly with Hulme on this, while recognising that the process by which a 'more constructive and imaginative engagement' with climate change might be achieved is a real challenge. A lived experience lens, for example, draws attention to the current realpolitik where 'win-sacrifice' is a grounded reality in everyday lives. In this context, recognising and accommodating different perspectives must play a major part in resolving the tensions that exist in 'win-sacrifice', as Jaeger et al. (2012) also advocate. Thus, a start to achieving a 'more constructive and imaginative engagement' needs, in our view, a more modest aim. This is: to shift the fundamental frame to one that lowers the stakes from fighting a battle/war and provides a calmer space for genuine political (at all scales) engagement about meaningful action on a big problem (but not a war).

Note also the common slogan:

> War is politics continued by other means,

Which we read as stating that war is another means of taking the politics out of decisions, a means that seeks justification in the urgency of the issue that is being decided. In Chap. 7 we examine further the possibilities for a new 'political' framing of the climate change challenge, a framing that involves human engagement through lived experience.

References

Camus, A. (1953, 1962). *The Rebel*. (Translated from the French by Anthony Bower; Foreword by Herbert Read). London: Hamish Hamilton/Peregrine Books.

Carbon Footprint Ltd. (2015). http://www.carbonfootprint.com (Accessed 23 February 2015).

Mastrandrea, M. D., Field, C. B., Stocker, T. F., Edenhofer, O., Ebi, K. L., Frame, D. J., Held, H., Kriegler, E., Mach, K. J., Matschoss, P. R., Plattner, G. -K., Yohe, G. W., Zwiers, F. W. (2010). *Guidance Note for Lead Authors of the IPCC Fifth Assessment Report on Consistent Treatment of Uncertainties*. Intergovernmental Panel on Climate Change (IPCC). Available at http://www.ipcc.ch.

Geoff, O' Brien, & O'Keefe, P. (2014). *Managing adaptation to climate risk: Beyond fragmented responses*. London: Routledge.
Harmer, N., & Rahman, S. (2014). Climate change response at the farm level: A review of farmers' awareness and adaptation strategies in developing countries. *Geography Compass*, *8*(11), 808–822. Doi:10.1111/gec3.12180.
Hasselman, K., & Voinov, A. (2012). The actor-driven dynamics of decarbonisation (Chap. 7). In C. Jaeger, K. Hasselmann, G. Leipold, D. Mangalagiu, J. D. Tabara (Eds.), *Reframing the problem of climate change: From zero sum game to win-win solutions*. London: Earthscan/Routledge.
Hegerl, G. C., Zwiers, F. W., Braconnot, P., Gillett, N. P., Luo, Y., Marengo Orsini, J. A., Nicholls, N., Penner J. E., & Stott, P. A. (2007). Understanding and attributing climate change: The physical science basis. In S. Solomon, D. Qin, M. Manning, Z. Chen, M. Marquis, K. B. Averyt, M. Tignor & H. L. Miller (Eds.), *Contribution of working group I to the fourth assessment report of the intergovernmental panel on climate change*. Cambridge, United Kingdom and New York, NY, USA: Cambridge University Press.
Hewitt, T. & Smyth, I. (1992). Is the world overpopulated? In T. Allen & A. Thomas (Eds.), *Poverty and development in the 1990s*. Oxford: Oxford University Press in association with The Open University.
Hobbs, J. (2012). Putting people at the heart of the climate debate (Chap. 6). In C. Jaeger, K. Hasselmann, G. Leipold, D. Mangalagiu, J. D. Tabara (Eds.), *Reframing the problem of climate change: From zero sum game to win-win solutions*. London: Earthscan/Routledge.
Hulme, Mike. (2009). *Why we disagree about climate change*. Cambridge: Cambridge University Press.
Hulme, Mike. (2013). *Exploring climate change in science and society: an anthology of Mike Hulme's writings, interviews and speeches*. London: Earthscan/Routledge.
IPCC. (2007a). *Climate change 2007: Synthesis report*. Cambridge, United Kingdom and New York, NY, USA: Cambridge University Press.
IPCC. (2007b). *Climate change 2007: Summary for policymakers*. Cambridge, United Kingdom and New York, NY, USA: Cambridge University Press.
IPCC. (2014a). Summary for policy makers. In C. B. Field, V. R. Barros, D. J. Dokken, K. J. Mach, M. D. Mastrandrea, T. E. Bilir, M. Chatterjee et al. (Eds.), *Climate change 2014: Impacts, adaptation and vulnerability*. Contribution of Working Group II to the Fifth Assessment Report of the Intergovernmental Panel on Climate Change. Cambridge, United Kingdom and New York, NY, USA: Cambridge University Press, pp. 1–32.
IPCC. (2014b). Summary for Policy Makers. In O. Edenhofer, R. Pichs-Madruga, Y. Sokona, E. Farahani, S. Kadner, K. Seyboth et al. (Eds.), *Climate change 2014: Mitigation of climate change*. Contribution of Working Group III to the Fifth Assessment Report of the Intergovernmental Panel on Climate Change. Cambridge, United Kingdom and New York, NY, USA: Cambridge University Press, pp. 1–32.
Jaeger, Carlo, Hasselmann, K., Leipold, G., Mangalagiu, D., & Tabara, J. D. (Eds.). (2012). *Reframing the problem of climate change: from zero sum game to win-win solutions*. London: Earthscan/Routledge.
Kempton, W. (1991). Public understanding of global warming. *Society and Natural Resources*, *4*(4), 331–345.
Mangalagiu, D., & Cameron, J. 2012. Challenges and opportunities in financing the response to climate change (Chap. 10). In C. Jaeger, K. Hasselmann, G. Leipold, D. Mangalagiu, J. D. Tabara (Eds.), *Reframing the problem of climate change: from zero sum game to win-win solutions*. London: Earthscan/Routledge.
Nyukuri, E. (2013). *Climate change, policy and vulnerability of tropical forest communities in East Africa*. Ph.D. Thesis, Milton Keynes, The Open University.
Sen, Amartya K. (1999). *Development as freedom*. Oxford: Oxford University Press.
Stern, Nicholas. (2010). *Blueprint for a safer planet: How we can save the world and create prosperity*. London: Vintage Books.
Toulmin, Camilla. (2009). *Climate change in Africa*. London: Zed Books.

Chapter 7
Lived Experience and Engagement on Climate Change

Abstract Within the dominant overarching 'threat and war' frame of climate change are nested many specific frames. One of these concerns the public will to know and act on climate change which, itself' has three broad problem dimensions that both overlap with each other and are partly oppositional. First, there is a perceived problem of communication of climate science which in earlier formulations could be addressed through more effective, targeted dissemination. More recently, however, it has been argued that deliberative processes of engagement between scientific and lay actors are required. Second, there is the nature of multiple perspectives that lead to disagreement on climate change and the question of how such disagreement may be put to the productive use. Third, is the resistance to engaging with disturbing information, except in an abstract way, through social norms of attention that shape what we think and talk about, and which lead to socially organised denial of climate change. Social norms of attention may be seen as placing limiting epistemological boundaries around lived experiences and hence a challenge for the fundamental argument of this book that views them as a positive resource.

7.1 Introduction

In the final paragraphs of Chap. 6 we examined briefly attempts to change an enduring frame of climate change intervention from 'win–lose' to 'win–win', through seeking, for example, co-benefits with other major societal challenges and changing to sustainable, prosperous green economies (Jaeger et al. 2012). This is the kind of polycentric framing that is advocated by Hulme (2009, 2013a). While win–win appears as a major advance from 'win–lose' or 'zero sum game', we also pointed out, that the reframing as conceived by Jaeger et al. (but not Hulme) still appeared to accept, and to take place within, an underlying frame of climate change as 'threat and war'. This frame requires sacrifice in the present, consequently leading to a certain degree of contradiction in the idea of 'win–win'.

D. Abbott and G. Wilson, *The Lived Experience of Climate Change*,
DOI 10.1007/978-3-319-17945-2_7

The threat on which we must intervene (or wage war) on climate change through mitigation, adaptation (or a mixture of the two) concerns the predicted largely negative biophysical and human impacts. Often, the impacts take on a deterministic flavour, for example the oft-repeated claims that climate change could result in real wars involving physical conflict and/or refugees. The range of intervention necessary to win the climate change war is very broad indeed, from international agreements to national policies to local, including individual, actions.

A key point for this chapter is that frames themselves appear at different levels and one may have frames within frames, subject to rules of consistency and non-contradiction at the different levels. Thus within the broad frame of 'threat and war' nest many specific frames where climate change is a problem of, for example:

- Political will. For instance, Jaeger et al. (2012) claim that 'political leaders who made climate change a central plank of their agenda [in the past] are [now] slowly distancing themselves'. Later in the same chapter they argue, 'Indeed, it is only through participation in the innovative governance initiatives of the innumerable regional and bottom-up actions that we can gain the necessary experience and create the political conditions needed for a global solution'.
- Market failure, where Stern (2010: 11) famously wrote:

> When we emit greenhouse gases we damage the prospects of others … Markets then fail in the sense that their main coordinating mechanism - prices – gives the wrong signals. That is, prices – of petrol or of aluminium produced with dirty energy, for example – do not represent the true cost to society of producing and using those goods… Greenhouse gas emissions constitute the greatest market failure the world has seen.

'Market failure' then justifies national and internationally coordinated action to correct for the failure, by providing an agreed means to put a price on greenhouse gas emissions.

- International development that is based on energy-intensive industrialisation. For example, in 1998, six years after the first World Summit on Environment and Development in Rio de Janeiro, Carley and Spapens wrote of China: 'Its expressed willingness to work with the global community in reducing carbon emissions notwithstanding, China remains adamant that economic development at home should not suffer to solve a problem it regards as largely created by the industrialised world' (Carley and Spapens 1998: 38).
- The public will to know and act, which includes the challenge of effective communication of, and engagement with, the climate change problem, but which has also begun to search for deeper meaning around the processes that are involved (see below).
- The enduring idea of human superiority and domination over nature, which Midgley (2014: 122, 123) claims is a source of climate denial. This idea started to take shape as a product of scientific progress attributed to 'The Renaissance' period in Europe between the 14th and 17th centuries. It became amplified with the onset of the industrial revolution that began in Britain at the end of the 18th century and soon spread through Europe and North America. In the prelude to

> her claim regarding climate deniers, Midgley (Ibid: 119) quotes a thinker of the Renaissance—Francis Bacon (of 'Knowledge itself is power' fame whom we also cited in Chap. 4)—as stating that we should 'conquer and subdue her [i.e. nature], to shake her to her foundations'.

Of these specific problem frames, the public will to know and act over climate change concerns us most in this book and chapter. Within the overarching frame it represents the public will to understand the threat and wage the good war against it. Meanwhile the public will to know and act itself has several dimensions which structure the sections that follow. We start in Sect. 7.2 with ideas around communication of climate change (especially the science), and then move in Sect. 7.3 to (lack of) engagement with the challenge and the multi-faceted nature of disagreement over it. We end the chapter with the concept of socially organised denial and the role of lived experience in facilitating or hindering engagement around climate change (Sect. 7.4).

7.2 The Public Will to Know and Act as a Problem of Science Communication

The advocates of polycentric, win–win (re)framings that we have cited in Sect. 7.1 above are, or have worked as, scientists. Mike Hulme's Ph.D. and the period following concerned the application of quantitative, especially statistical, models derived from observational climate data (Hulme 2009: xxx). Jaeger's et al.'s (2012) edited collection stems from papers presented at the international conference 'Action for Climate—beyond the zero sum game' of the Global Climate Forum in Barcelona, April 2010. They make no secret of their pedigree: 'It [the book] is written by scientists... ' and 'our general approach is that of the scientist' albeit from both natural science and social science disciplines (Ibid: Kindle location 267).

It is reasonably safe to assume that the polycentric, win–win framings of Hulme and Jaeger et al. arise as a response to deeply embedded problems of communicating the science of climate change, the seriousness with which we should take that science, and the multiple, often competing, perspectives on the subject. They are also linked therefore to the 'public will to know and act' frame where, of its several sub-frames, the original and still the largest body of work focuses on issues of poor communication by scientists to policy makers and the general public. This results in turn in failure to act significantly at any scale, even though the natural science states with a high degree of certainty what the bio-physical impacts are likely to be, and the social science the socio-economic impacts (Moser and Dilling 2007a; Whitmarsh et al. 2011a). For scientists, it is a frustrating experience (Fig. 7.1).

Thus in Moser and Dilling's edited collection that draws from the United States experience, we find a concerted attempt to move beyond simply attempting to disseminate information and knowledge in traditional ways (Dilling and

Fig. 7.1 The frustration of communicating climate change to the public. (Cartoon by Jon Kudelka. *Source* http://www.smithsonianmag.com/science-nature/talking-about-climate-change-how-weve-failed-and-how-we-can-fix-it-180951070/?no-ist.)

Moser 2007). Instead, contributors in Part I of their book argue for more effective, targeted dissemination, which include:

- Better knowing the recipient audience and how it perceives climate change, and tailoring the fit between the problem and existing beliefs (Bostrom and Lashof 2007; Leierowitz 2007; Moser 2007; Chess and Johnson 2007).
- Learning from how effective advertising gets its message across through cultural resonance (Ungar 2007) and more effective use of mass media. (Dunwoody 2007).
- Supporting scientists to get their message across (Cole and Watrous 2007).

While some of these contributions have a slightly distasteful whiff of manipulation about them in order to communicate the message, others in Part I provide case studies of actions that have sought to listen better to local citizens (Pratt and Rabkin 2007), and to integrate climate change with the ‘real concerns, constraints and strengths’ that exist in many poor communities (Agyeman et al. 2007). As a further extension which attempts to move beyond simple but effective dissemination, two chapters concern engagement between different actors, including scientists and non-scientists, to create dialogue, ‘new ways of talking’ (McNeeley and Huntington 2007) and a ‘new kind of conversation’ (Regan 2007). Dialogue is here seen as a space where contrasting understandings and perspectives may be reconciled.

Moser and Dilling (2007a) are also concerned that communication and engagement should stimulate changes in individual and group behaviour, and indeed the more fundamental social change that they believe is required to meet the climate challenge. In their concluding chapter (Ibid: 2007b) they argue that 'motivators' and overcoming the 'mental barriers' to change which appear in the contributions throughout the book are complementary and both are needed for communication to be effective, that is to facilitate a desired social change (Ibid: Kindle location 6447). They add, however, that 'our mental models or habits of thought are among the most critical barriers to change' (Ibid: Kindle location 6454). They then further identify a thread of thinking that runs through the volume that calls for a shift from one-way message delivery to 'more engaging, dialogic forms of communicating' and call for the need for broad public dialogue (Ibid: Kindle location 6542). This does not apply only, however, to communication between scientists and non-scientists as perceived mostly by McNeeley and Huntington, and Regan cited above (Ibid: Kindle location 6577). In addition, they note developments in the United States concerning engagement across common societal divides (for example, between richer and poorer communities), which they characterise as 'boundary crossing' to find common ground and common language (Ibid: Kindle location 6694). Overall, they state that their book contributors understand the communication process as an intricate and multifaceted form of mutual engagement (Ibid: Kindle location 6512).

Whitmarsh et al.'s (2011a) edited collection has similar concerns to those of Moser and Dilling. The contributors derive mostly from the UK, but with Australia, the United States and Canada also represented. The volume shows a determined shift towards what one contributor calls a 'participatory and collegial approach to knowledge production' (Leith 2011: 114). The editors and some contributors also consider their mission to go beyond knowledge co-production to include 'civic and community forms of engagement [that] offer an expanded role for individuals in respect of defining climate change responses and shaping social change, and also conferring potential benefits in terms of fostering self-efficacy, democracy, community cohesion and social inclusion' (Whitmarsh et al. 2011b: 272).

Substantively, the collections of Moser and Dilling (2007a) and Whitmarsh et al. (2011a) are both rich with case studies and insights, while being somewhat eclectic in their treatment of climate change communication and engagement. This is, of course, to be expected in edited volumes with many contributors. The subject matter ranges from how to get the message across to lay publics (drawing on a variety of practices and techniques, several of which derive from social psychology) to discussion of constructivist forms of engagement and communication which lead to collaborative efforts in joint knowledge production and beyond. In the latter, Whitmarsh et al. go further than Moser and Dilling, but again this is to be expected as four years separate the two volumes, a period during which a 'deliberative turn' (providing the conditions for reasoned debate on climate change and reflexivity—see Sect. 7.3 below) was increasingly established.

While it is dangerous to summarise Moser and Dilling (2007a) and Whitmarsh et al. (2011a) in one sentence, for the purposes of our book we suggest that, whether referring to practices and techniques for better dissemination of the message, or to deliberative processes of engagement, both are substantively concerned with 'smart communication' where no one size fits all. This is even true, despite Whitmarsh et al.'s claim of potential further social benefits of collaborative deliberation, social benefits that are, alas, aspirational rather than real. As one contribution in Whitmarsh et al. notes, 'Tensions between purely instrumental understandings of citizen engagement as a means to cut carbon emissions and normative democratic understandings of empowering citizens without prescribing the outcomes of their engagement remain unaddressed (Hoppner and Whitmarsh 2011: 52).

Finally, in this section, it is noteworthy that a more recent consideration of these tensions (Corner and Groves 2014) has conceptualised them as being between the norms that govern scientific practice (disinterested, neutral and objective) and those that govern the social domain (shaped by values, ideology and social identity). Corner and Groves (Ibid.) then argue that the way forward lies in 'creating and supporting new institutions and societal infrastructure that provide a buffer between the science of climate change and the complex challenge of engaging the public'. These 'new hybrid institutions' would involve scientists and 'actors from across the social and political spectrum', whose purpose would be to 'catalyse new conversations about climate change… to allow people to express and discuss their concerns, fears, dreams and hopes for the future, providing answers to that troubling question 'how shall we live?' This again begs the question, however, of how such institutions might be created, especially when they challenge powerful interests such as that of science and scientists in public institutions (see Chap. 4, Sect. 4.4) where knowledge is guarded and manipulated in order to legitimise claims to authority within specific arenas. In this regard they remain aspirational.

7.3 The Public Will to Know and Act, and Disagreement Over Climate Change

The above edited collections stem from the clear observation that 'there is something in *how* [emphasis in original] we communicate climate change that is failing' (Moser and Dilling 2007a: Kindle location 216). Further literature in the genre starts from roughly the same point, but with new insights. Thus, Hulme (2013b: 252–264) adds from his UK experience (but which applies to other affluent countries) the falling confidence over the years in many public institutions and professionals, including but not only, scientists. This lack of trust was heightened especially in late 2009/early 2010 after the release of emails from, to and between climate scientists at the Climatic Research Unit at the University of East Anglia in the UK which suggested a lack of transparency in presenting research results (popularly known as the 'Climategate' scandal, also referred to briefly in Chaps. 3 and 4). It was further exacerbated during the same period after errors were found in the 2007 IPCC Assessment Report, notably in the Report of Working Group (WG) II, 'Impacts, Adaptation and vulnerability' (see Chap. 6 for our analysis of this report). The most publicised of these errors concerned exaggerated claims about the rate of decline of Himalayan glaciers and the percentage of land in the Netherlands that lie below sea level, the latter prompting the Dutch Parliament to call for an independent line-by-line review of the entire WGII report.

Partly arising from these insights and observations of Hulme, and from critical analysis of the deliberative turn noted above, our second sub-frame of the public will to know and act encompasses the all too apparent, diverse competing perspectives on climate change, the nature of disagreement, and even the productive potential of disagreement. Our analysis centres on Hulme (2009, 2013a) again and a book from a political science perspective by Machin (2013).

The title of Hulme's (2009) book—'Why we disagree about climate change'—signals its intention directly. Disagreement is ubiquitous and inherent to climate change. Hulme (2009) argues that the proximate causes of disagreement are (page numbers are in brackets):

- Our different understandings of the relationship of scientific evidence to 'other things'—essentially the other items in this list (106).
- The different values we place on our responsibilities to future generations, humans and nature, and climate risks (139).
- Our different belief systems—from materialist to ethical beliefs about reality (142–177).
- Our different ways of understanding and perceiving the risks associated with climate change (207).
- The different ways that climate change is framed and communicated and the ways it is filtered through the 'intuitive world views of those listening' (245).
- Our different ways of relating climate change to other pressing global issues of poverty and vulnerability, of population, and of 'excessive affluence' in a minority of countries. In other words, our different ways of relating climate change to global justice and (in)equity (281).
- Different (self-)interests across scales—from affluent communities protesting against wind farms to poorer countries defending their right to development. Also the different approaches to negotiating the politics—from bottom-up and participatory to top-down and hierarchist. Overall, 'the quagmire of global [including climate] politics' (317–320).

Hulme (2009) also argues that climate is more than a physical phenomenon that has become a problem, but 'an idea that binds together the physical world and our cultural imagination' (Ibid: 32). That is, we give meaning to the physical world of climate through our culture, our ways of life. Thus, in a broader sense, the sources of our disagreement about climate change 'lie deep within us, in our values and in our sense of identity and purpose' (Ibid: 364).

There are intuitive resonances here between culture and our (the authors of this book) notion of lived experience. Hulme, however, never explicitly explains his own understanding of culture, neither in his 2009 book nor his 2013 anthology, despite repeated references to the concept. The omission is unfortunate because, without clear explanation, misunderstanding is likely. This is especially so, as, over 60 years ago, Kroeber and Kluckhohn (1952) compiled a list of 164 definitions of "culture".

In order to aid engagement with Hulme's arguments we provide our definition as follows. It is a synthesis of several other definitions in existence.

> Culture is an integrated and shared system of learned values, beliefs and behaviour patterns that are characteristic of the members of any given society, institution or organisation and which gives shared meaning to the phenomena we encounter in our lives. Culture refers to the total way of life of particular groups of people. It includes everything that a group of people thinks, says, does and makes, its customs, language, material artifacts and shared systems of attitudes and feelings. Culture is learned and transmitted from generation to generation.

Reading between the lines of Hulme, he probably wouldn't have difficulty with the above. Let us now examine how it might relate to the lived experience of climate change through two examples that we have covered in Chap. 1—the UK flood victims of late 2013/early 2014 and the displaced East African forest dwellers.

Example 1: Culture and the UK flood victims

If you have shared values such as a strong work ethic, if you believe that through working hard you obtain your just rewards, if you carry that belief into your behaviour and are rewarded with a big house and lovely family in a countryside setting, but are forced suddenly one day to abandon your home because of a severe weather event, your experience could be that you have been treated unjustly by a supreme creator or the Earth Goddess Gaia or simply fate. 'Why us? It's unfair', is a probable refrain.

In this—partly hypothetical—example we note a link between culture (values, beliefs behaviour) and lived experience. The experience of the extreme weather event, however, also influences in turn potentially one's values, beliefs and behaviours. For example, your shared experience with others who suffered from the flood may re-ignite the value of community, or of pulling together to save a business as we saw in Chap. 1. The relationship of lived experience with culture, therefore, is circular.

The lived experience is also more than what might be placed within the notion of culture (at least our definition of it). You may have lost your livelihood. You may feel vulnerable because you are from a poor background without the material resources and capabilities to rebuild your livelihood. You might be the opposite. You also will experience commentary and explanations from outsiders, especially meteorologists and politicians. You may feel powerless and out of control as you have to deal with insurance companies and various authorities. You will undoubtedly engage with others and reflect together on what has happened and how to respond. Also undoubtedly, you will reflect privately to try and make sense of your predicament and the choices open to you. All of these are part of the dynamism of lived experience as we developed it in Chap. 1.

Example 2: Culture and vulnerability of East African forest communities

If you place a high value on the forests and believe them to represent your identity, if you believe the products they supply have health benefits, if you are, however, displaced from the forest through political machinations and land-grabbers for economic gain, if the displacement is exacerbated by climate mitigation interventions concerning forest conservation, your experience may be of loss, physically but also mentally in terms of identity. Overall your already existing sense of vulnerability and disempowerment will be exacerbated. Again, however, your experience of displacement requires more than the concept of culture to explain it. The whole lived experience is also influenced by your structural poverty, your illiteracy and your powerlessness to do anything about it. In the language of Amartya Sen (Chap. 6) you lack choices. You are not, however, completely disempowered and lacking agency as you may make alliances with other indigenous groups across the world to fight your corner, engaging in joint reflection with them.

Hulme himself often elides cultural with social explanations. In an essay published in his 2013 anthology he states categorically that climate change 'has become as much a social phenomenon as much as a physical one' (Hulme 2010a). Our book is no place, however, to enter discussions about how many angels can dance on the head of a pin. There is significant overlap and accord between ourselves and Hulme. We are broadly on the same wavelength, and that matters more than continuing to pick over what he means by culture. For the purposes of our book we have a strong preference for lived experience as a better explanatory and analytical concept because it is understood as a form of knowledge that may be compared with scientific and other expert/professional knowledges and with other lived experiences. The content of lived experience varies from individual to individual, but the idea of it does not—it is more grounded than culture in the everyday practice of our human attributes of reflexivity and communication.

Illuminating his core idea about disagreement being inherent to climate change, Hulme (2010a) states: 'The arguments about the causes and consequences of climate change—and the solutions to it—have become nothing less than arguments about some of the most intractable social, ethical and political disputes of our era: the endurance of chronic poverty in a world of riches; the nature of the social contract between state and citizen; the cultural authority of scientific knowledge; and the role of technology in delivering social goods.'

Further sources of disagreement concern the ideological baggage that Hulme states is carried by climate change, for example to support 'ideologies of racism, the human mastery of nature, the sanctity of a pristine nature, and the preference for stability over change' (Hulme 2009: 32).

Are such deep-seated disagreements the reason for the lack of public will to know and act? Are they therefore a source of despair? Not for Hulme. He states that 'disagreeing is a form of learning' (Ibid: xxxiv). Because of our disagreements about climate change, we have to reveal these deeper reasons 'rather than pretending that louder, crisper and slicker communication of science will somehow bully the world to a convergence'(Hulme 2007). This is not, however, a prognosis for despair: 'It is only once we truly understand how deep our differences are, and respect them—differences in beliefs, values, goals, instruments, politics—that we will be in a position to think more clearly about what we really want to happen in the future' (Ibid.).He is also quoted as saying, 'Climate change can only be understood from a position of dissensus, rather than artificially solved by creating consensus' (Blackman 2009).

These sentiments chime with our own writings about treating difference as a positive resource for new knowledge rather than despair (Abbott and Wilson 2012). Note again subtle differences in terminology to that used by Hulme. We contend that difference is the source of social learning (see Chap. 5). We learn new things through our differences, not by being the same. Disagreement is in our view an intermediate product of difference—we disagree because of our differences. This matters only in the sense that, while our differences *might* cause us to disagree, we don't *have to* do so in order to learn. We may learn additionally through observation of the behaviours of others who are different from ourselves. We may also learn through trying to understand their mindsets and finding common ground and ways of accommodating one another, none of which *necessarily* requires disagreement.

Hulme's analysis of the productive potential of disagreement leads him to argue for a reframing of the goals of climate change intervention to seek co-benefits with other social problems (Hulme 2010b). This is the polycentric framing which we have discussed above and in Chap. 6. He also calls for a 'new pragmatism' and engagement between actors for a 'set of mini-lateral approaches' to intervention, so, rather than trying to get everybody to agree, we should 'make multiple tracks through smaller coalitions of interests' (2010c). This approach, which we again support, is in stark contrast to the system of trying to reach big, all-encompassing agreements through the United Nations.

Building on Hulme's arguments, and arguing from a political and theoretical perspective, Machin (2013) explicitly engages in 'celebrating disagreement' (the phrase appears in the title to Chap. 6 of her book) as a basis for reaching decisions in relation to climate change and other pressing social issues. Thus, she argues that having an environmental consciousness 'creates ground of disagreement' rather than producing a unified set of ecological values. Our disagreements may then be used to 'revitalise democratic politics' (Ibid: 5). This is because, 'it is precisely through democratic disagreement in the political realm that citizens become inspired, engaged and interested in an issue' (Ibid: 27, 85). She is again in line with Hulme when she states that there is 'no one correct position from which to assess climate change' (Ibid: 88) and that any decision between alternative courses of action is by definition underpinned by disagreement (Ibid: 101). Hers is thus a plea for making politics the central focus of climate change debates, not the science or economics or the science-culture nexus.

Along the way she exposes the weaknesses of other mechanisms that have been proposed for meeting the challenge of climate change. She draws them together under four headings: new technologies and market mechanisms; ethical consumerism; green republicanism; and deliberative democracy. She then describes her preferred mechanism under the label of radical democracy. We summarise her critiques and her advocacy of radical democracy below, together with our view on each (page numbers in brackets are to her book). Thus:

- New technologies and market mechanisms for putting a price on carbon have a place, but only if there is a 'real engagement with the people who are affected' (26). Our view: Agreed—this echoes our argument that lived experiences of those affected by climate change interventions in the present are needed in order to legitimise them (Chaps. 5 and 6).
- The ethical consumer who buys environmentally friendly ('green') and fairly traded products is engaged in a guilt-assuaging exercise through the 'individualisation of responsibility' (29, 32)—which is a flight from the politics that are required for necessary structural transformations and institutional change. Our view: This rhetoric tends to view the individual ethical consumer in a static way. Individualistic actions demonstrate both possibility and potential, and at the same time the inadequacy of the individualistic action by itself. Such actions are potentially able to create awareness of the wider politics that are needed.
- Green republicanism is a form of direct democracy. It recognises the need for collective political action at the community level by informed and committed citizens who are aware of their responsibilities as well as their rights and who arrive at rational agreement for 'the common good', which for greens is sustainability. However, Machin asks, who decides on the common good and are those who disagree excluded? Moreover, are others excluded from active participation in the community politics and reasoning over the common good because they have other burdens that effectively prevent them from playing a full role? Here, Machin provides a gendered example of differentiated burdens: 'While the green republican citizen participates in debates about the duties of citizenship, someone is minding the children, taking out the recycling and washing the reusable nappies' (63). Our view: The difficulties attached to green republicanism are well articulated, but is not the debate academic anyway? Direct democracy such as this does not exist anywhere in the contemporary world except possibly in a very limited sense at community levels, where the important decisions are made at higher (usually national) scales.
- Deliberative democracy requires reasoned debate between actors. It welcomes and respects difference but maintains that articulation of ideas in public and the requirement to give good reasons for one's position during a debate force reflexivity among participants (76). Thus deliberative democracy is a process of social learning that eventually leads to a consensual agreement with respect to what should be done. However, Machin (80) argues that deliberative democracy is not as free and equal to all participants as it claims. Citing Iris Young, she claims that the style of deliberation is always culturally specific and marginalises other ways of communicating, which silences or devalues the speech of some people (80). More generally, and now citing Diana Coole, she argues that the structural power relations that exist between different societal groups—poor and rich, women and men for example—affect the ability to 'participate freely, equally and rationally' (81). Our view: While we can agree with Machin's

critique, and of those whom she cites, the argument seems again to be too static and predictive about what *might* occur in the spaces that *attempt* deliberative democracy. We emphasise the word 'might' because power and powerlessness are not necessarily locked into a never-changing relationship. We emphasise also the word 'attempt' because Machin and others are arguing against deliberative democracy in its ideal form of human engagement. As with all ideal forms it simply never exists in practice. In other words, critiquing ideal forms tends to be akin to setting up straw people whom it is easy to knock down. Of course one cannot simply do away with power relations between people and social groups through apparently rational debate. Yet these are never situations of absolute power and absolute powerlessness. Power has different sources—for example, financial and material power, formal power between owner and worker, power of different forms of knowledge, power of alliance with other like-minded people—and there is always some room to negotiate the relations, if not to transcend them. Spaces that attempt deliberative democracy will contain some element of this negotiation, even if that is not the intention of the ideal. It might be sometimes that the power relations remain intractable but to assume that this will always be the case is wrong. It leads, moreover, to despair that anything may ever be transcended.

- Having found problems with the above forms of engagement and decision-making, Machin arrives at her own preference: radical democracy. She starts by attacking the idea of reaching consensus which she claims underpins the other forms: 'Contrary to dominant understanding, decision is underpinned not by consensus but by disagreement, for without a choice between real alternatives there can be no decision' (89). She cites Jacques Rancière who claims that consensus is the disappearance of politics (90). The only alternative is exclusion of those who disagree from the consensus which 'results in the eruption of antagonism in a heightened extremism that shifts from the political realm towards moral territory' (97). She then applies this analysis to climate change deniers whose confrontation with the scientific consensus becomes visualised as a moral one between good and evil, where the denier is perceived as an enemy who must be destroyed (Ibid.). The essence of radical democracy, however, is that it celebrates and thrives on disagreement to provide real choices where ultimately a decision is made between them. This paradoxically requires a consensus about the democratic process that allows for disagreement, choices and decision. Machin sums up the general idea in Mouffe's phrase 'conflictual consensus' where adversaries are 'legitimate adversaries rather than enemies to be destroyed' (92). She then summarises the argument as follows: 'In a democracy the decision that is eventually (but never finally) made is one that not all participants agree with, but that they all agree *to go along* with, at least until it can be challenged and the issue re-decided (her emphases)' (102). Here she echoes other literature which encapsulates 'agreeing to go along with a decision' as a deal that incorporates 'accommodation of interests' (Isaacs 1993). For democratic legitimacy, therefore, no substantive agreement on the decision is needed.

It is needed, however, on the importance of democracy. In this way, political opponents 'agree to abide by the decisions that are taken' (102).

Our view: We can agree with much of Machin's critique of the other mechanisms for reaching appropriate decisions in relation to climate change, although we have reservations concerning some of the points and claims she makes above. This last favoured mechanism—radical democracy—then leaves us with an abstract notion of creating a space to celebrate disagreement for provisional decisions that might be 'green'. Radical democracy in relation to decisions about climate change, however, requires a more thorough questioning.

Machin's starting point as summarised above is to claim that the idea and practice of consensus leads inevitably to the exclusion of those who disagree. Of course it may be that, and probably often is. Is it, however, that always? There is an issue of moving from what is potentially, and undoubtedly is sometimes, the case, to something that inherently is always the case. For the latter to be established as a matter of fact, one needs robust empirical evidence, which is in short supply in her book.

Climate change provides one domain where a dominant scientific consensus as created by the IPCC excludes 'deniers'. Machin, and others she cites in her analysis above, then generalises to refer to consensus as exclusion and the disappearance of politics. Is this really the case? Is not a more accurate depiction that, far from disappearing, politics is carried on by other means by the excluded? This depiction seems to fit well the tactics of the climate change deniers as exemplified by Hulme's analysis of the Climategate scandal and other attempts to discredit the scientific consensus that we have described above.

Moreover, surely questions may be asked about the exercise of radical politics where all parties are supposed to accept the decision out of respect for the democratic process even if they disagree with that decision? We make seven points with respect to this:

1. Is radical politics realistic, especially when it involves strongly felt views or where one fundamentally disagrees with the decision? We saw in Chap. 5 Leroi's (2014: 316) hardly complimentary depiction of direct, participatory democracy in 4th century BC Athens as 'squalid… institutionalised mob rule where demagogues roused the rabble'.
2. Who designs the supreme democratic process to which everyone will ultimately be subordinate and can it ever be 'perfect' so that dissenters to a decision will never complain about it?
3. Are those who focus on ends—that is those who have a strong vision—included in the radical democracy even though ultimately its democratic process may become, for them, subordinate to pursuance of that vision? Is not vision important and its pursuance through alliances and other ways of marshalling power a normal part of political process?
4. Does one have to sign up and agree/be part of a consensus to this political process and its own 'tyranny' in order to be a legitimate actor?

5. Shouldn't we also investigate how disagreement is currently used in climate change politics, but in a negative way—to neutralise or cut off sources of power such as scientific knowledge? We saw in Chap. 3 how disagreements over science neutralise each other and are used as such by political actors. Here scientific knowledge is a source of power, but so is scientific disagreement to neutralise this source of power.
6. Questions concerning how the decision is reached and who owns (and takes/has responsibility for) the decision reflect power relations among the actors surely, in radical democracy as elsewhere? In short does not the critique concerning social power that Machin makes so forcefully in relation to green republicanism and deliberative democracy apply also to radical democracy?
7. How can disagreement be made productive? This question concerns the micro-dynamics of narrative exchange at the level of the actors concerned. Radical democracy as expounded by Machin is silent on it. Surely some conditions need to be in place for any narrative exchange to be useful, including understanding and respecting the views of others even when you disagree. Habermas (who Machin critiques in her chapter on deliberative democracy) suggests that a 'background consensus' (we can't get away ultimately from a consensus somewhere!) of shared meanings lies behind all narrative exchange (quoted in Fischer 2003: 199). One of the authors of this book has noted the value to narrative exchange of shared experience of doing things together (Wilson 2008) and building trust between actors (Wilson 2006) where shared meanings are not automatically present (see also Chap. 8). None of these claims to be an easy process, mainly because of structural power relations, but if conditions for facilitating narrative exchange do not exist even in imperfect form, how are engagement processes—agreeing or disagreeing—to be made productive?

In summary, Machin presents us with a single big idea: celebrating disagreement through radical politics to reach decisions on climate change. The text is normative throughout, arguing towards radical democracy in terms of what 'should be'. It is far removed from the 'what is' of the present. This is fine as far as it goes, but without any indications of how one might make the transition, it is ultimately unconvincing.

Thus, Machin's book raises for us as many questions as it answers. The sum of these questions is that the positive case for celebrating disagreement is not adequately made. While accepting many of the critiques, neither is the case for demonising consensus. Perhaps it would be more fruitful to consider both as having productive potential, depending on context.

7.4 The Public Will to Know and Act, Socially Organised Denial and Lived Experience

We have met in Sects. 7.2 and 7.3 some fundamental challenges that the 'Public will to know and act' frame raises. They include on one hand challenges of communicating the message and, prior to that, challenges of agreeing what the message is or even that there is only a single over-riding message. On the other hand there is the challenge of recognising, understanding and accepting disagreement on climate change, and putting such disagreement to productive use.

It would be reasonable to hypothesise that analysis of lived experiences of the public would throw some light on these challenges concerning its will to know and act. Lived experience, however, rarely features in any of the literature that we have reviewed to date in this chapter. Machin (2013: 114) alludes to it when she states, 'Our perception of our environment is bound up with our understanding of who we are, and who we understand ourselves to be affects our perceptions of our environments'—and hence our disagreements about climate change and the environment. In his anthology, Hulme (2013a: 93) refers to 'what climate change means to people and the relationships between people and places over time'. Later (Ibid: 188) he states that the global temperature and the target of holding its increase to two degrees 'has no resonance with the everyday experience of weather and climate… it has difficulty engaging the human imagination'.

One of the works published in Hulme's (2013a) anthology is a review of an anthropological study by Norgaard (2011) of a small town community in Norway, titled 'Living in denial: climate change, emotions and everyday life'. As signalled by the words 'everyday life', lived experience is central to the analysis, where an epistemological and ontological sub-frame emerges that examines the ways in which the hegemonic power of particular ideas and the quest for ontological stability in everyday lives put boundaries around what is permissible to attend to in thought and conversation. This, in relation to climate change, can lead to what the author describes as socially organised denial of the problem and the loss of the ability (and hence power) to set public agendas in relation to it.

Here we spend some effort analysing Norgaard's study because hers is a powerful lens through which to view lived experience in the context of the frame of the 'public will to know and act'. It is also highly relevant to our own analysis of lived experience in the chapters which follow. (Kindle location numbers to her book are placed in brackets).

Norgaard lays out her stall in the book's Prologue (137): 'This book is about how people experience disturbing information about climate change. It tells a story about what goes on behind the scenes to create the public face of apathy. It is a story that uses the voices of members of one small town… It is a story that I hope will help us understand the complexity of the lived experience of people around the world as we struggle collectively to make sense of this significant problem'.

Citing Eviatar Zerbave (195, 306, 316), she points out that society teaches us what to pay attention to and what to ignore. That is, we learn 'cognitive traditions' through a process of socialisation. Thus, we think not only as individuals and as human beings but also as members of particular communities with certain distinctive cognitive traditions that affect the way we process the world in our minds. These 'cognitive traditions' emerge from shared everyday life practices over time.

In this language, the relevance of climate change to daily life is a function of socially shaped systems where cognitive traditions and thought communities shape how and whether groups of people think about the topic and whether they perceive it as relevant (205, 208). They help create an ordered world that provides 'ontological stability', a concept which Norgaard attributes to the British sociologist, Anthony Giddens. Ontological stability represents a world where reactions to climate change are quite uniform among residents and are clearly controlled by social norms of behaviour that also reinforce residents' sense of themselves as good people (466, 477). In this way the maintenance of ontological stability tends to normalise knowledge about climate change rather than act upon it (2491).

Norgaard then introduces the related concept—also attributed to Giddens—'ontological security', which refers to the confidence that most human beings have in the continuity of their self-identity and the constancy of the surrounding social and material environments of action (1212): 'Merely thinking about climate change raises a series of questions related to ontological security: What will Norwegian winters be like without snow? What will happen to farms in the community in the next generation? Will they… begin growing oranges in Norway'? (It's better not to think too hard about these things.)

The 'social organisation of denial' about a phenomenon such as climate change is, for Norgaard, a function of the social shaping of our awareness, memories, and thought patterns (211). It emphasises that ignoring occurs in response to social circumstances and is carried out through a process of social interaction with others who share the same cognitive tradition (255). It is the process by which individuals collectively distance themselves from information because of 'norms of emotion, conversation, and attention' (259).

We, the authors of this book, should stress at this point that this is not usually denial in the sense of outright rejection of information but 'the failure to integrate this knowledge into everyday life or to transform it into social action' (272). Here we may note the crucial fault line with the 'information deficit' model of science communication of climate change where the public fails to respond because of a lack of information (Chap. 10). Instead we have a 'social organization of denial' where the public on a collective level actively resists available information (298). In other words the public does not have the will to know and thus it normalises, or minimises, disturbing information about climate change in such a way that denial feels to people like everyday life (1757). Information on the topic is known in an abstract sense but not integrated into the sense of immediate reality (800). Norgaard posits this as a fundamental challenge for an enlightened, democratic, and modern society whose basic premise is that information (especially scientific information) will lead to concern and response on the part of the public and public institutions (1110).

How are particular social systems of memory or cognition created and maintained? In answering this question, Norgaard uses the word 'enforced' rather than 'maintained', presumably because her answer involves the concept of power. Drawing on the work of Antonio Gramsci and his emphasis on how social control is enacted through the acceptance of ideas that prevent social change (217), she points to the structuring of everyday reality through social, political and economic institutions (1915) from local to national scales and through the particular context of one of the world's richest countries, Norway, where oil and gas have played a significant part in generating that wealth (261). This rich fossil fuel endowment induces emotions of guilt among citizens as they perceive themselves as being responsible for climate change while much poorer people in other parts of the world suffer the human consequences. The resulting social cognition of denial is then played out through 'social norms of attention—that is, the social standard of "normal" things to think about', which are 'powerful, albeit largely invisible social forces shaping what we actually do think about' (1621).

Climate change and its politics for this Norwegian community are not, therefore, normal things to attend to. For Norgaard, social norms of attention represent a subtle maintenance/enforcement of power which 'works in part by robbing the powerless of the inclination or ability to develop their own interpretations of political issues… Without the power to determine what sorts of questions are worth discussing in public, citizens are deprived of an important power, the power to define what is worthy of public debate, what is important' (820, 824). Feelings of guilt, moreover, are submerged through construction of positive self-representations or at least acceptance of community and individual citizen roles 'in the existing order of things because they can see or imagine no alternative to it, or because they see it as natural or unchangeable' (1932).

Another inevitable factor for this small community that influences the norms of attention is an acute sense of the local aligned to a dislike of change. Multifaceted aspects of the local—'dialect, community, insiders and outsiders, sense of time and place, tradition'—create norms of attention which make 'events that happen elsewhere seem distant and less real' (452). Extending this to lived experience (local or not), Norgaard quotes Eliasoph: 'The way people make sense of everyday experiences usually discourages them from thinking thoughts that might challenge the status quo'… People come to hold 'a mistaken interpretation of how the world actually is' (1937, 1939, 1998, 232).

If all of the above applies to a single small community in Norway, what about the rest of the world? Norgaard generalises her analysis as a citizen of the United States: 'What we all can observe in the public silence on climate change in the United States and around the world is not in most cases a rejection of information per se, but the failure to integrate this knowledge into everyday life or to transform it into social action' (277, 800).

Norgaard's analysis is impressive and powerful. It adds significantly to our understanding of the role of lived experience in the failure of the public will to know and act on climate change. It is also an uncomfortable analysis for a central argument of our book that lived experience may be put to productive use for public action and policy on climate change (Chaps. 8–10). Norgaard provides a start in helping us to escape from our discomfort when she recognises that 'conversations' with others 'can help people understand their relationships to the larger world', as well as obscure them (811) which seems to be happening in the small community she studied. She also points out that failure to act on climate change is not a result of personal inhumanity or greed. Rather, socially organised denial, is a response to our underlying 'human capacity for empathy, compassion and an underlying sense of moral imperative to respond, even as we fail to do so' (938).

This leads us to Chap. 8, whose main aim is to explore further how we might, through engagement with others, transcend the boundaries of socially organised denial in relation to lived experience and other forms of knowledge.

References

Abbott, D., & Wilson, G. (2012). The lived experience of climate change: Complementing the natural and social sciences for knowledge, policy and action. *International Journal of Climate Change Impacts and Responses, 3*(4), 99–114.

Agyeman, J., Doppelt, B., Lynn, K., & Hatic, H. (2007). The climate-justice link: Communicating risk with low-income and minority audiences (Chap. 7). In S. Moser & L. Dilling (Eds.), *Creating a climate for change: Communicating climate change and facilitating social change*. Cambridge: Cambridge University Press.

Blackman, S. (2009). Top British boffin: time to ditch the climate consensus (Chap. 43). In *Exploring climate change in science and society: An anthology of Mike Hulme's writings, interviews and speeches*. London: Earthscan/Routledge.

Bostrom, A., & Lashof, D. (2007). Weather and climate change (Chap. 1). In S. Moser & L. Dilling (Eds.), *Creating a climate for change: Communicating climate change and facilitating social change*. Cambridge: Cambridge University Press.

Carley, M., & Spapens, P. (1998). *Sharing the world: Sustainable living and global equity in the 21st century*. London: Earthscan.

Chess, C., & Johnson, B. (2007). Information is not enough (Chap. 14). In S. Moser & L. Dilling (Eds.), *Creating a climate for change: Communicating climate change and facilitating social change*. Cambridge: Cambridge University Press.

Cole, N., & Watrous, S. (2007). Across the great divide: Supporting scientists as effective messengers in the public sphere (Chap. 11). In S. Moser & L. Dilling (Eds.), *Creating a climate for change: Communicating climate change and facilitating social change*. Cambridge: Cambridge University Press.

Corner, A., & Groves, C. (2014). Breaking the climate change communication deadlock. *Nature Climate Change, 4*(9), 743–745.

Dilling, L., & Moser, S. (2007). Introduction. In S. Moser & L. Dilling (Eds.), *Creating a climate for change: Communicating climate change and facilitating social change*. Cambridge: Cambridge University Press.

Dunwoody, S. (2007). The challenge of trying to make a difference using media messages (Chap. 5). In S. Moser & L. Dilling (Eds.), *Creating a climate for change: Communicating climate change and facilitating social change*. Cambridge: Cambridge University Press.

Fischer, F. (2003). *Reframing public policy: Discursive politics and deliberative practices*. Oxford: Oxford University Press.

Hoppner, C., & Whitmarsh, L. (2011). Public engagement in climate action: Policy and public expectations (Chap. 3). In L. Whitmarsh, I. Lorenzoni & S. O'Neill (Eds.), *Engaging the public with climate change: Behaviour change and communication*. London: Earthscan.

Hulme, M. (2007). Climate change: From issue to magnifier (Chap. 53). In *Exploring climate change in science and society: An anthology of Mike Hulme's writings, interviews and speeches*. London: Earthscan/Routledge.

Hulme, M. (2009). *Why we disagree about climate change*. Cambridge: Cambridge University Press.

Hulme, M. (2010a). Heated debate (Chap. 41). In *Exploring climate change in science and society: An anthology of Mike Hulme's writings, interviews and speeches*. London: Earthscan/Routledge.

Hulme, M. (2010b). The year climate science was redefined (Chap. 49). In *Exploring climate change in science and society: An anthology of Mike Hulme's writings, interviews and speeches*. London: Earthscan/Routledge.

Hulme, M. (2010c). Climategate, scientific controversy and the politics of climate change (Chap. 47). In *Exploring climate change in science and society: An anthology of Mike Hulme's writings, interviews and speeches*. London: Earthscan/Routledge.

Hulme, M. (2013a). *Exploring climate change in science and society: An anthology of Mike Hulme's writings, interviews and speeches*. London: Earthscan/Routledge.

Hulme, M. (2013b). After climategate… never the same. In *Exploring climate change in science and society: An anthology of Mike Hulme's writings, interviews and speeches*. London: Earthscan/Routledge.

Isaacs, W. (1993). Taking flight: Dialogue. *Collective Thinking and Organizational Learning, Organizational Dynamics, 22*, 24–39.

Jaeger, C., Hasselmann, K., Leipold, G., Mangalagiu, D., & Tabara, J. D. (Eds.). (2012). *Reframing the problem of climate change: From zero sum game to win–win solutions*. London: Earthscan/Routledge.

Kroeber, A., & Kluckhohn, C. (1952). Culture: A critical review of concepts and definitions. *Papers Peabody Museum of Archaeology and Ethnology, 47*(1), viii, 223. Harvard: Harvard University.

Leierowitz, A. (2007). Communicating the risks of global warming: American risk perceptions, affective images, and interpretive communities (Chap. 2). In S. Moser & L. Dilling (Eds.), *Creating a climate for change: Communicating climate change and facilitating social change*. Cambridge: Cambridge University Press.

Leith, P. (2011). Public engagement with climate adaptation (Chap. 6). In L. Whitmarsh, I. Lorenzoni & S. O'Neill (Eds.), *Engaging the public with climate change: Behaviour change and communication*. London: Earthscan.

Leroi, A. M. (2014). *The lagoon: How Aristotle invented science*. London: Bloomsbury Circus.

Machin, A. (2013). *Negotiating climate change: Radical democracy and the illusion of consensus*. London: Zed Books.

McNeeley, S., & Huntington, O. (2007). Postcards from the (not so) frozen North: Talking about climate change in Alaska (Chap. 8). In S. Moser & L. Dilling (Eds.), *Creating a climate for change: Communicating climate change and facilitating social change*. Cambridge: Cambridge University Press.

Midgley, M. (2014). *Are you an illusion?*. Durham: Acumen.

Moser, S. (2007). More bad news: The risk of neglecting emotional responses to climate change information (Chap. 3). In S. Moser & L. Dilling (Eds.), *Creating a climate for change: Communicating climate change and facilitating social change*. Cambridge: Cambridge University Press.

Moser, S., & Dilling, L. (Eds.). (2007a). *Creating a climate for change: Communicating climate change and facilitating social change*. Cambridge: Cambridge University Press.

Moser, S., & Dilling, L. (2007b). Toward the social tipping point: Creating a climate for change (Chap. 32). In S. Moser & L. Dilling (Eds.), *Creating a climate for change: Communicating climate change and facilitating social change*. Cambridge: Cambridge University Press.

Norgaard, K. M. (2011). *Living in denial: Climate change, emotions and everyday life*. Cambridge, Massachusetts: The MIT Press.

Pratt, L., & Rabkin, S. (2007). Listening to the audience: San Diego hones its communication strategy by soliciting residents' views (Chap. 6). In S. Moser & L. Dilling (Eds.), *Creating a climate for change: Communicating climate change and facilitating social change*. Cambridge: Cambridge University Press.

Regan, K. (2007). A role for dialogue in communication about climate change (Chap. 13). In S. Moser & L. Dilling (Eds.), *Creating a climate for change: Communicating climate change and facilitating social change*. Cambridge: Cambridge University Press.

Stern, N. (2010). *Blueprint for a safer planet: How we can save the world and create prosperity*. London: Vintage Books.

Ungar, S. (2007). Public scares: Changing the issue culture (Chap. 4). In S. Moser & L. Dilling (Eds.), *Creating a climate for change: Communicating climate change and facilitating social change*. Cambridge: Cambridge University Press.

Whitmarsh, L., Lorenzoni, I., & O'Neill, S. (Eds.). (2011a). *Engaging the public with climate change: Behaviour change and communication*. London: Earthscan.

Whitmarsh, L., Lorenzoni, I., & O'Neill, S. (2011b). Conclusion: What have we learnt and where do we go from here? In L. Whitmarsh, I. Lorenzoni & S. O'Neill (Eds.), *Engaging the public with climate change: Behaviour change and communication*. London: Earthscan.

Wilson, G. (2006). Beyond the technocrat? The professional expert in development practice. *Development and Change, 37*(3), 501–523.

Wilson, G. (2008). Our knowledge ourselves: Engineers (re)thinking technology in development. *Journal of International Development, 20*, 739–750.

Chapter 8
Lived Experience, Science and a Social Imagination

Abstract It is important to expand the boundaries imposed by social norms of attention around lived experiences in order to generate a broader and deeper knowledge about pressing social issues, such as climate change, and thereby regain power to set public agendas. For similar reasons it is equally important in academia to establish interdisciplinary endeavours between the scientific (natural and social) disciplines. A combination of the hermeneutic and emancipatory knowledge interests that are elaborated by Jürgen Habermas, and the sociological (which we reframe as 'social') imagination of C. Wright Mills are two complementary expressions of this broader and deeper knowledge. In either case, it develops through purposeful transboundary engagement of actors, that is, engagement between diverse lived experiences, between the different social and natural science disciplines, and between lived experiences and the scientific disciplines. Productive transboundary engagement is a challenge, however, requiring mutual trust, shared experience, continual reinforcement, structured discussion with trigger topics and facilitation.

8.1 Introduction

We, the authors, have friends! One of them moved a few years ago to a small town in the English Midlands, sufficiently removed from the big cities to have its own character. He tells us that it is the friendliest place that he has ever known. He greets everyone he meets with, 'Hello, how are you doing, are you all right?' and they greet back with the same (the question part of the greeting being redundant conversation). He chats for ages with his neighbour opposite about this and that.

The place where our friend lives appears to have similarities with the small town in Norway that was the subject of anthropological study by Norgaard (2011) and which we covered in Chap. 7. It is a bit larger (population about 19,000 compared to 10–14,000 in the Norwegian town) and almost certainly it is not so isolated. It is full of 'good', decent people, however, who go about their daily

D. Abbott and G. Wilson, *The Lived Experience of Climate Change*,
DOI 10.1007/978-3-319-17945-2_8

lives, greet everybody and chat to those whom they are more acquainted, just as Norgaard found in Norway.

If Norgaard, or even we, were to study our friend's community from an anthropological perspective, we are sure that she/we would also soon find norms of attention of what is acceptable to attend to in conversation and what is not. These norms in turn would generate socially organised denial—a collective resistance to unsettling information about a major social issue—and serve to preserve ontological stability/security among the townspeople, that is, a strong, enduring sense of who they are, of their being.

This is what it means to be a friendly community. There is a sense of belonging and mutual trust where collective adherence to the norms of attention may be assumed. It enables one in turn to engage in conversation with virtually anybody else in the community. This, we believe, is different from having friends. With friends you are more likely to explore, if only occasionally, beyond the state of the local roads, the closing down of a much loved local shop, the weather that day or week. Thus, the closer that you are to friends the more that you feel safe to explode the norms of attention and reveal deeper concerns. Note here our use of the word 'safe'. To unpack 'norms of attention' a little, they almost certainly mean what is *safe* to attend to as much as what is *acceptable*.

Norms of attention lead to 'conversation norms', another term used by Norgaard when she writes: 'Yet conversations are also governed by rules called "conversation norms" that shape what is acceptable and not acceptable to say in different contexts' (Ibid: Kindle location 1416). Strictly speaking, we have been referring above to this latter term. Conversation norms within a community seem to constitute, or at least overlap strongly with, Habermas' idea of 'background consensus' which we also came across briefly in Chap. 7, and which is the system of shared meanings that he claims lies behind all narrative exchange (Fischer 2003: 199). By extension of our analysis above, background consensus indeed limits narrative exchange to safe topics, but we further suggest that friends in conversation may sometimes breach its boundaries or otherwise expand it to contain a wider range of topics.

The reason that friends may do this is the trust that has built between them over a significant length of time. It is trust in one another as individual people, rather than as part of a community, that enables one to say things that would be unthinkable in a 'friendly' conversation. We return to the idea of trust and its enabling function for expanded forms of engagement below.

For the moment, we explore norms of attention/conservation in the context of another feature of many small communities—cultural homogeneity. Norgaard does not spell out the possible diversity in her small Norwegian town, particularly its ethnic and religious diversity. Her Chap. 1, however, constitutes an introduction to life in the town, with the opening sentence proclaiming: 'Culture and everyday life organise people's experience of our world (including larger global events such as climate change) in different ways' (Norgaard 2011: Kindle location 302). In this chapter, she highlights shared culture as: the strong sense of community and social networks that go back generations, symbolic traditions, a regionally specific dialect, a strong sense of insider-outsider and a proper way to do things that is still strongly influenced by the Lutheran Christian Church. Overall, in her town there exists cultural homogeneity which reinforces, and is reinforced by, social norms of attention and the consequential conversational norms.

The small English town where our friend lives has elements of the above, but probably not to the same depth. We, the book authors, are even further away from the cultural reality that Norgaard attributes to her small Norwegian town. We live respectively in a large town (population about 150,000) and a bigger city (population about 310,000). Our everyday reality is much more heterogeneous and multi-cultural. Our lifestyles are different in other ways too. We engage with others at work and on our many trips abroad with people from all over the world. Moreover, many of our friends are widely dispersed. Do social norms of attention and conversational norms have the same meaning for us? Do we have more possibilities to transcend them and engage in conversations about other things, especially political problems? Even if we do have those possibilities do we make use of them? In 1759, the French social critic and philosopher Francois Voltaire wrote his satirical tale 'Candide' where the hero decides that the best thing he can do in a world of misery and inhumanity is 'tend his own garden' and live peacefully with his neighbours (Voltaire 1759/2005).

Friendships also depend on being relaxed with one another, allowing for serendipity and having a laugh together. While deeper engagement about issues does occur, as noted above, it is still occasional. Trying to engage too often in 'heavy' conversations about climate change and other pressing global issues might be the best way to lose our friends, or at least they will tell us to 'shut up'.

At the end of Chap. 7, we quoted Norgaard (Kindle location 811): 'Conversation can help people understand their relationships to the larger world or obscure them'. Conversations of course expand beyond our 'communities' and our friends. We might happen to hold a conversation with whoever we meet in the wider world, including those we work with, those we meet on our travels including leisure travels, and even those we meet virtually via social media. They might include *inter alia* climate scientists, climate activists, climate deniers and climate agnostics, but here the cognitive boundaries are more likely to be characterised by our assumptions about one another. It is also important to note that everyone we meet in the 'wider world' is likely to have norms of attention that are diverse and different from our own. Will the conversations we hold with one another constitute a clash of norms of attention and be uncomfortable? Will we try and avoid such clashes and search for shared interests in order to be even more restricted and cautious in what we talk about? Will there be an opportunity to go beyond clashing, and enrich one another?

This brings us to the central questions that we address in this chapter:

- To what extent is it desirable to expand the boundaries of our lived experiences beyond prevailing social norms of attention?
- To what extent is it desirable to expand the scientific knowledges about climate change beyond their boundaries?
- If it is desirable to do so, how might such expansions be facilitated and who might facilitate it?

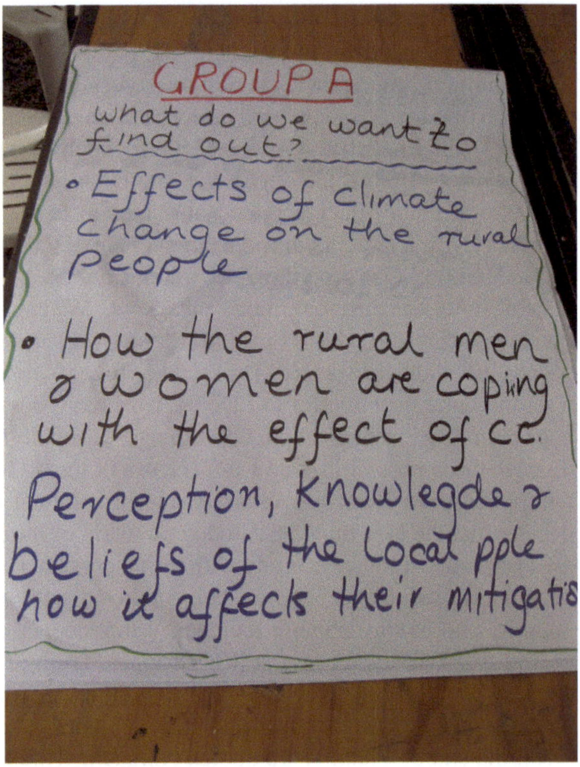

Fig. 8.1 First page of a flip-chart presentation from a group of NGO practitioners in Uganda who have been deliberating about lived experience and climate change

Figure 8.1 is an example of deliberations from a group of NGO practitioners in Uganda, which provisionally suggests positive answers to these questions before we start to explore them in the sections below.

8.2 To What Extent Is It Desirable to Expand the Boundaries of Our Lived Experiences Beyond Prevailing Social Norms of Attention?

The main reason for expanding the boundaries of our lived experiences is to generate a broader and deeper knowledge, that is, a more complete knowledge, of the world and the great social challenges it faces, including of course the challenge of climate change. Through this broader and deeper knowledge, we may then regain the power to set public agendas (Norgaard 2011: Kindle location 1934; see also Chap. 7). Our book argues that we generate this knowledge by crossing epistemological boundaries to share and compare experiences with others who may be very different from us. It is a deeper, more complete knowledge that we generate within ourselves and share with others, thus creating a virtuous circle of knowledge generation—or so the idea runs.

The idea of sharing knowledge to generate a deeper, more complete knowledge is common to several academic disciplines. For example, economics as a discipline is far removed from the anthropology of Norgaard, but Thomas Picketty in his best selling book about economic inequality—'Capital in the twenty-first century'—states:

> Knowledge and skill diffusion is the key to overall productivity growth as well as the reduction of inequality both within and between countries. We see this at present in the advances made by a number of previously poor countries, led by China. These emergent economies are now in the process of catching up with the advanced ones. By adopting the modes of production of the rich countries and acquiring skills comparable to those found elsewhere, the less developed countries have leapt forward in productivity and increased their national incomes. The technological process may be abetted by open borders for trade, but it is fundamentally a process of the diffusion and sharing of knowledge – the public good par excellence – rather than a market mechanism. (Picketty 2014: 21).

There is no doubting the discipline in the above quotation. The language virtually 'breathes' economics, the 'dismal science' as it is often called because of its many doom-laden predictions, although not in this quotation from Picketty. On the contrary, he is extolling the virtue of open borders of knowledge that allow for diffusion and sharing. His reference point is the macro-scale of national economics, but the same message may be found at local scales. Chipika and Wilson (2006) found for example, among small-scale light engineering firms in Zimbabwe, that those whose networks extended beyond kinship ties were more innovative than those who were restricted to family.

Knowledge is well established in the discipline of economics because it is needed for innovation in response to market signals. We quoted in Chap. 5 a definition of innovation as being 'knowledge for productive use' (Chataway 2005). The basic logic is that market competition drives the need both to do things better (improve quality and productivity for example) and do better things (make new products or improved adaptations of existing ones) (Forbes and Wield 2002).

Moving from economics to life, and lived experience, we also expand our knowledge through sharing with others, learning from and with them. This is the dynamic core of our conceptualisation of lived experience that we created in Chap. 1. By analogy with our local economics example above of small-scale light engineering firms in Zimbabwe, however, context is important. If our networks are restricted to family or to a geographical community in which we live, social norms of attention and conversational norms limit what we may engage about. This is what Norgaard (cited above) means when she states that conversation may help people understand their relationship to the wider world, but may also obscure it. In other words, it depends with whom you engage and talk, and the subjects you talk about. We explore in Sect. 8.4 the practical possibilities for shifting the balance towards understanding one's relationships with the wider world and away from obscurity. For the moment, we simply hypothesise that engagement with others who are not bounded by the same norms of attention as us is key to obtaining and assimilating an expanded knowledge of climate change.

What, however, do we mean by expanding our knowledge? We referred above to broadening and deepening. *Broadening* is essential. It refers to understanding a wider spectrum of knowledge that speaks to climate change debates. For example it might mean, inter alia, understanding more about:

- The physical science and its complexities
- The macro-debates concerning the impacts on the world economy and what this would mean for both poor and rich countries; likewise the differentiated social impacts
- The social relations of transferring to a low-carbon economy, a major goal of the European Union to be reached by the year 2050 (European Commission 2011; see also Chap. 3). This means, among other things, switching from a mainly fossil-fuel-based energy industry to one based on renewables (wind, waves, photovoltaics). We do not argue with this as a goal, but there is scant literature on the process of transformation beyond stating new skills that will be needed (an EU agenda is 'new skills and jobs'—European Commission 2010). What will happen to people who work in the current energy industry? Will they all be re-skilled and nobody made redundant? What will happen to the linkages to other areas—for example, the manufacturers of central heating systems that are based on fossil-fuel boilers? What will happen to the many small businesses that have contracts to maintain these systems and other energy appliances in buildings, especially if they become so well insulated and draught-proof that they have less need of heating? We could go on, but you surely acknowledge the message. These are not frivolous questions.
- The political processes of reaching (or failing to reach) decisions on intervention measures, from international to nation scales.
- The relationship between politics and markets. Do we try and generate the conditions so that capitalist markets and competition will enforce green innovation and solve the climate change problem for us, or do we attempt to govern capitalistic markets more directly?
- Human relations with 'nature'. For example, how do we react when land is turned over to biofuels, or when conservation of tropical forests increases the vulnerability of already vulnerable forest communities, and a renewable energy installation spoils the 'peace and quiet' of an affluent community in Europe and elsewhere?
- The linkages to other pressing global issues—for example, poverty and inequality in all its facets, wars, terrorism, international development.

Deepening is qualitatively different from broadening the range of inter-related topics about which we should know. Here we examine two complementary approaches: (i) communicative action, and hermeneutic and emancipatory learning; (ii) social imagination.

8.2.1 Communicative Action and Hermeneutic and Emancipatory Learning[1]

When Norgaard acknowledges that conversation may help people understand their relationships to the wider world she draws explicitly on the idea of communicative action as developed by the German philosopher, Jürgen Habermas.

Habermas' theory of communicative action resonates strongly with the idea of social learning (as outlined in Box 5.3, Chap. 5), and with knowledge sharing as implied above. He starts his journey with the simple but fundamental question: What is the primary distinguishing feature that makes us human? His answer: our capacity to reflect on what we do and learn. Habermas then argues that this fundamental capacity, which is present in all human beings, is the source of two capabilities (Edgar 2006: 62–64):

- Our ability to 'labour', by which Habermas means our ability to transform our physical environment or 'nature' for productive use.
- Our ability to interact and communicate with each other, not just in the sense of conveying information, but to justify our reflections in the form of discussion, debate and challenge.

Habermas is certainly not the first person to refer explicitly to the first of these capabilities. In the nineteenth century, for example, the ability to labour formed the underlying assumption in the work of Karl Marx. Habermas argues, however, that by itself our ability to labour reduces human ability for reflection and creativity to a *technical* interest of knowledge and communication. Important as this is, it ignores the human ability to (a) reflect on society, and interpret it through discussion, challenge and debate with others (i.e. what he calls a *hermeneutic interest* of knowledge); and (b) challenge and shape that society (an *emancipatory interest* of knowledge). Habermas' division of knowledge into technical, hermeneutic and emancipatory interests is often the most-quoted aspect of his work (Habermas 1987a). Box 8.1 explains further, but briefly, these knowledge interests of human beings:

Box 8.1 Technical, hermeneutic and emancipatory knowledge interests

Habermas posits three interests of knowledge. These are summarised by Mohan and Wilson (2005) as:

Technical—aimed at the material reproduction of society and how one controls and manipulates one's environment (instrumentalism). This is largely based on *scientific* approaches.

Hermeneutic—aimed at enhancing understanding and transforming consciousness through binding consensual norms. This is largely based on

[1]Section 8.2.1 is adapted from a book, 'Learning for development', that one of the authors co-wrote with Hazel Johnson (Johnson and Wilson 2009: 20–26).

approaches whereby communication and intersubjectivity allow a greater *understanding* of meanings of the social worlds we inhabit.

Emancipatory—aimed at breaking free from structures which limit our options and which have hitherto been regarded as beyond human control. This involves critical self-*reflection* so that a person can truly recognise the source of their problems.

Hermeneutic learning by itself equates to the ways in which we might broaden our knowledge on climate change, as outlined above. The combination of hermeneutic and emancipatory learning, however, represents for us a *deeper* form of knowledge production.

The difficulty is that on a day-to-day basis there is a self-perpetuating tendency to privilege our technical knowledge interests that constitute exchanges of information of how to do things better in a taken-for-granted capitalist world. Habermas (1984: 143–242) has grappled with this tendency where what he calls *instrumental rationality* gains precedence over *communicative rationality* which might breed challenge to the social system (see Box 8.2 and also Borda Rodriguez 2008: 42–43). He does not, however, reject instrumental rationality as such, recognising that it is essential for societies to function. For him, and for us, the key issue is the way that instrumental rationality has a tendency to build on itself with the evolution of society, and is in danger of eclipsing communicative rationality—what he calls 'colonisation of the lifeworld' (Ibid.; Habermas 1987b). This does not bode well for social learning and deeper knowledge generation that is emancipatory, which comes about through discussion, debate and challenge—in short a process which he calls *communicative action* (Ibid.) (Box 8.2).

Box 8.2 Habermas' forms of rationality

Instrumental rationality is summarised by Edgar (2006: 74) as the rational choice for the achievement of any given end. It is thus a rationality associated with selection of means rather than of ends. Instrumental rationality is fundamental, for example, to the application of technology, judged by its efficiency and effectiveness, as well as in social administration and in the formation of social and economic policies.

Communicative rationality and communicative action. In contrast to instrumental rationality, communicative rationality concerns the rational choice of ends, arrived at by communicative action—free and open discussion of all relevant persons without any form of coercion (Edgar 2006: 23).

Thus how and why learning and knowledge generation occur cannot be divorced by the kind of learning and the knowledge it produces, which leads to the concern below (Sect. 8.4) about *how* to promote deeper learning and knowledge that is connected to climate change. For Habermas, however, deeper knowledge about climate change would be linked to a wider vision—of political emancipation and social justice to be achieved through communicative action. More broadly, still he seeks to contribute towards what he terms the unfinished project of modernity, where reason prevails (Edgar 2006: 96–100).

In communicative action, one always has to give reasons for one's statements (a rationality for seeking consensus that is challenged by Machin 2013: 80–86). In this, Habermas' argument is similar to that of the Brazilian educationalist, Freire, and his book *The pedagogy of the oppressed* (1972) which sought through open discussion and debate to arrive at the roots of oppression and challenge them. It is within these wider visions that Hulme (2009, 2013) and others seek to locate climate change.

Habermas is well aware of the tension between instrumental and communicative rationality, as indicated above. There are, however, two further issues that arise, even when space is allowed for open and transparent discussion in intervention on any social challenge, including that of climate change.

Both issues stem from Habermas' assertion of a pre-condition for communicative exchange where the participants are able to draw on assumptions that are held in common, mutually drawing on a shared *lifeworld* (Habermas 1984, 1987b, 1990—see Box 8.3). This he has also described, noted in Sect. 8.1, as the 'background consensus that lies behind all narrative exchange' (quoted in Fischer 2003: 1999). When considering intervention related to climate change, however, such a background consensus is often likely not to exist a priori between the different stakeholders and therefore communicative action cannot establish itself successfully. The first issue then becomes whether (or not) a common background consensus can be generated? Section 8.4 explores this issue more thoroughly.

Box 8.3 Lifeworld

This is the stock of competences and knowledge that people use to negotiate their way through everyday life, to interact with others, and ultimately to create and maintain social relationships (Edgar 2006: 89). Our lifeworld competences make communicative action possible (see Box 8.2). They typically are used to maintain our relationships with people we know, not only through taken-for-granted agreement, but also through challenge and negotiation. The interactions involved, however, require that we establish a basic shared view of the way the world works through establishing common meanings and common interpretations of the lifeworld. (Ibid: 90). In other words, we establish a *shared* lifeworld or background consensus.

The second issue concerns where a background consensus might with justification indeed be said to exist between stakeholders (for example, because they are all bound by the same norms of attention, or are all climate scientists, or all climate economists). Do not the common assumptions that form the background consensus then limit the possibilities for emancipatory/transformative learning as the participants are trapped within a shared lifeworld whose existence appears, to them, natural and therefore not open to challenge?

The two issues represent the tension for this chapter, between commonality and difference that we explore again in Sect. 8.4. Commonality between individuals and groups enables a background consensus that in turn allows for communicative exchange as a basis for shared learning and knowledge generation. In addition, commonality provides a secure base for joint organisation and action with potential for learning and innovation. In contrast, under conditions of difference, a background consensus and joint organisation and action are more difficult. It is difference, not commonality, however, that is ultimately the source of learning and deeper, new knowledge.

8.2.2 The Social Imagination

Another way of thinking about deepening knowledge about climate change through engagement with others concerns developing and assimilating a 'social[2] imagination'. This way of thinking derives from the classic work by C. Wright Mills, 'The sociological imagination' (Mills 1959/2000) whom we first met in Chap. 2 (Sect 2.3 and Box 2.4). He explains as follows (page numbers in brackets refer to the 2000 edition):

> [What people] need, is a quality of mind that will help them to use information and to develop reason in order to achieve lucid summations of what is going on in the world and of what may be happening within themselves… The sociological imagination enables its possessor to understand the larger historical scene in terms of its meaning for the inner life… (5).
>
> The sociological imagination enables us to grasp history and biography [that is, the personal story that each of us lives out] and the relations of the two within society. That is its task and promise (6)… it is by means of the sociological imagination that men[3] now hope to grasp what is going on in the world, and to understand what is happening in themselves (6, 7).

Note in the above quotations, three themes that formed the subject matter of Chap. 7 and the present chapter in relation to climate change:

- Providing more and more information is not sufficient, and may be counter-productive unless accompanied by the capacity to develop reason
- What is needed is a transformation of information to understand the meaning of climate change in the context of the world in which we live, which corresponds

[2]Note our use of the adjective 'social'. Mills used the term 'sociological imagination', because his book is a polemic about social science and social scientists and what they should, and conversely should not, be and practice. He made greatest claim for the social science discipline of sociology. In our book, however, we are not limited to any disciplinary understanding, still less privileging one. The adjective 'social' captures perfectly well the kind of imagination of which we espouse.

[3]It is somewhat grating to read Mills' directly here and elsewhere when he refers to 'men', 'he', etc. to donate people and she/her. We seek your indulgence for a book that was written in the 1950s by a man.

to the Habermasian notions of hermeneutic and emancipatory knowledge interests as opposed to merely technical interests
- We need to assimilate internally deeper understanding of what is going on in the world into our lived experience, to develop that 'quality of mind', and indeed to combine it with that lived experience.

Thus to have a social imagination with respect to climate change is not simply to have more knowledge on the topic, but to deepen our understanding of that knowledge. It means that we are prepared to engage broadly and deeply with the topic in combination with other social issues, and to connect it to and embed that greater understanding within our lived experiences. In relation to the latter, Mills points out the limitations of daily experience that become exposed by possessing a social imagination which enables its possessor to:

> Take into account how individuals, in the welter of their daily experience, often become falsely conscious of their social positions... By such means the personal uneasiness of individuals is focused upon explicit troubles and the indifference of publics is transformed into involvement with public issues. The first fruit of this imagination... is the idea that the individual can understand his own experience and gauge his own fate by locating himself within his period, that he can know his chances in life only by becoming aware of those of all individuals in his circumstances. (5).

Climate change covers the globe. There is no hiding place for anyone. The main difference in terms of human intervention is that some of us have the relative 'freedom' (that is the resources and capabilities in Amartya Sen's logic—see Chap. 6) to adapt and even attempt some mitigation, while many others, through their structural circumstances, have very little room for manoeuvre. The social imagination thus enables us to become internally aware of not only those in similar circumstances, but also how our daily actions affect others in near and distant places.

The social imagination then enables us, according to Mills, to appreciate the impotence of individual action because of the social structures within which we live:

> Insofar as an economy is so arranged that slumps occur, the problem of unemployment becomes incapable of personal solution. Insofar as war is inherent in the nation-state system and in the uneven industrialisation of the world, the ordinary individual in his restricted milieu will be powerless... to solve the troubles this system or lack of system imposes upon him...
>
> To be aware of the idea of social structure and to use it with sensibility is to be capable of tracing such linkages among a great variety of milieux. To be able to do that is to possess a sociological imagination. (10).

It is a straightforward matter to paraphrase the first part of this citation in relation to climate change. Thus, insofar as economic life is so arranged that crises in nature occur, the problem of climate change and other environmental states (biodiversity, pollution, etc.) become 'incapable of personal solution'. Insofar as economic life is predicated on energy use, and that nobody or no institution is able to coordinate the anarchic arrangements of nation states, the individual is 'powerless' to do anything about it (11).

The words in the above paragraph that we have retained within quotation marks suggest to us, however, that Mills here is making too clean a separation between the individual and social structures. As we pointed out in our critique of some of the literature that we cited in Chap. 7, it is important to have a more dynamic, and consequently less static, view of personal action. Yes, individual action is no panacea for tackling the climate change challenge. Yes, it is limited in what it may achieve. This is not to state, however, that it achieves *nothing.* Neither is it to state that those who engage in individual actions are incapable of recognising their limitations, or of reflection on that recognition and so beginning to gain a 'quality of mind' in relation to climate change intervention. It is important to recognise that reflection and quality of mind develop potentially from practice as well as ideas.

The strength of the concept of 'social imagination', however, is the way it explicitly links understanding of the wider world and its problems to what we call lived experiences of people. The latter concept also helps ground Mills' notion of 'quality of mind', which to us is the assimilation of our greater understanding into our lived experience rather than just linking the two. Thus, rather than 'the failure to integrate this [wider] knowledge into everyday life or to transform it into social action' (Norgaard 2011: Kindle location 272), Mills presents us with a scenario where such integration has become successful. Rather than 'our mental models or habits of thought' being 'among the most critical barriers to change' (Moser and Dilling 2007a: Kindle location 6454), he presents us with the same scenario where prevailing mental models and habits have been overcome. (Note that both citations in this paragraph also appear in Chap. 7).

The overlaps of the concept of 'social imagination' with the Habermasian notions of hermeneutic and emancipatory knowledge interests are fairly obvious. In fact for Habermas and his followers, one would develop a social imagination through the conversational style of communicative action (see above), rather than expecting social scientists in the discipline of sociology to do the job on a largely ignorant public. Norgaard (2011: Kindle location 1409) makes the link explicitly. Using Habermas as one of her reference points she states:

> … when people get together and talk, a number of important things happen… Conversations can help people understand their relationships to the larger world… They can engage the sociological imagination.

This leads us to our second and third questions that concern, respectively, the disciplining features of the sciences (not only sociology but the social and natural sciences in general), and a closer analysis of the mechanisms and dynamics of human engagement in developing, or failing to develop, a social imagination.

8.3 To What Extent Is It Desirable to Expand Scientific Knowledges About Climate Change Beyond Their Boundaries?

We have alluded in previous chapters to the ways in which different science subjects might frame a topic like climate change, but that each is limited by the concepts which it uses for analysing. Thus, the physics of climate change use the laws of thermodynamics, the different wavelengths of radiant energy that we receive from the sun and the energy absorption properties at different wavelengths of certain atmospheric 'greenhouse' gases—notably carbon dioxide, methane, nitrous oxide and water vapour—to explain why our climate is warming. The first law of thermodynamics states that energy is neither created nor destroyed (but may be converted from one form to another), the second law that heat energy will only flow spontaneously from a hotter to a cooler body (for example, from the Sun to the Earth). Together they form the nuts and bolts of the physics framework. By putting these concepts to use we may explain global warming as an imbalance caused by carbon dioxide and other greenhouse gases trapping energy received from the sun. Further, physics concepts may also be applied to air and ocean circulations which help explain different climates across the world. None of this, however, explains the biophysical impacts on biodiversity and ecosystems involving plants and animals—that is the province of what are called the life sciences such as biology. None of it, moreover, explains the human impacts—that is the province of the social sciences and their concepts.

As we explored briefly in Chap. 3, it is no wonder that different sciences are called 'disciplines'—they discipline us into certain ways of thinking about problems and provide the conceptual tools for doing so. This is very useful for examining particular features of a problem such as climate change. Each is, however, limited by its particular disciplinary boundaries and cannot claim to provide a whole story of the phenomenon.

Mills (1959/2000) had several things to say about this. Early in his book he equates a social (sociological) imagination with something akin to what we might call today a multi-disciplinary imagination, that is, to move easily between disciplines and so overcome the limits posed by each:

> … that [sociological] imagination is the capacity to shift from one perspective to another – from the political to the psychological; from examination of a single family to comparative assessment of national budgets of the world; from the theological world to the military establishment; from considerations of an oil industry to studies of contemporary poetry. It is the capacity to range from the most impersonal and remote transformations to the most intimate features of the human self – and to see the relations between the two. (6).

Adapting these shifts across conceptual boundaries to, and grounding them in, contemporary climate change, we might paraphrase the above quotation as: from the intractability of reaching international political agreements to the psychology of mental barriers of change in individuals; from examination of a single family to comparative assessments of national carbon foot prints across the world; from what the religious texts tell us about past great floods and other extreme events to the analyses of the World Meteorological Organisation; from considerations of the energy industry in its totality (fossil fuels and renewables) to artistic depictions of wind turbines.

As the antithesis to his concept of the sociological imagination and the more holistic approach he implies above, Mills (14, 15) posits 'physical science', claiming: 'As an intellectual style, physical science is thought by many as somehow inadequate'. By 'intellectual style' we assume the way in which the conduct of physics abstracts and breaks matter into tinier and tinier components—thus the tiniest particle is first theorised, followed by empirical confirmation using a sophisticated piece of kit—and then to re-assemble them as reality. 'Men of science, it is widely felt', he writes, 'no longer try to picture reality as a whole or to present a true outline of human destiny'.

Writing about the favour among many social scientists of his day to try and develop 'grand theories' to explain social phenomena, just as the physical sciences do, he comments: 'They never, as grand theorists, get down from the higher generalities to problems in their historical and structural contexts. This absence of a firm sense of genuine problems makes for the unreality so noticeable in their pages' (Ibid: 32).

As well as arguing against grand theory in the social sciences, Mills critiques another approach that tends to remove the contexts of history and social structures from explanations of phenomena: 'abstracted empiricism' (Ibid: 50–75). In this form of social science research, the 'fundamental source of information is a sample of individuals. The questions asked in these studies are put in terms of the psychological reactions of individuals'. In other words, we see throughout the discipline of psychology similar 'atomisation' into small units as we find in physics.

Psychology of course has further atomised the individual with respect to its key concerns—the mind and the brain. Thus the philosopher Midgley (2014: 133) writes of the drift of psychology to study the mind 'by more "scientific" means, that is, by something that looks more like chemistry or physics', and (Ibid: 147) that 'reductivist' psychologists have decided that human life 'can henceforth be safely left to neurological experiments' that use modern brain-imaging methods.

Today, within climate change studies, we see the same tendency towards what Mills (1959/2000: 61, 67–68) calls 'psychologism' within the 'communication of the problem' literature that we analysed in Chap. 7. Several chapters in Moser and Dilling's (2007a) edited volume have this tendency. Much of it is summed up by the editors themselves as understanding the motivators and mental barriers to behavioural change on climate change (Moser and Dilling 2007b: Kindle location 6579)

Mills appears, therefore, to have fully understood the limiting tendencies of individual disciplines. We may capture the nature of what he appeared to be edging towards in 1959 through the idea of an *interdisciplinary imagination*, which we introduced in Chap. 4, Sect. 4.6. Rather than hopping from one perspective to another, as in multidisciplinarity, interdisciplinarity involves work at the epistemological boundaries between scientific disciplines to generate new insights. Mills himself alluded to interdisciplinarity when he argued that several prominent intellectuals of his day individually could be located within more than one social science discipline, for example, economics, political science, psychology, sociology:

> They are in and out of the social sciences and so they are all… They may of course specialise in one institutional order, but insofar as they grasp what is essential to it, they will come to understand its place within the total structure, and hence its relations with other institutional domains. For in considerable part, it is becoming clear its every reality consists of these relations. (Mills 1959/2000: 138, 139)

This, however, is all very well for the social sciences where individual disciplines have regularly 'strayed' over their boundaries into other disciplines. In the process, they have enriched and been enriched by fresh insights. Climate change knowledge, however, is also informed by the physical and life sciences. The former of these, especially, holds significant authority and hence power (Chap. 3). Working Group I of the IPCC has always comprised a study of the physical science basis of climate change (Chap. 6). In our view it needs bringing into the interdisciplinary endeavour, if for no other reason than to move beyond the authority provided by one discipline and to accept the polycentrism of knowledge on climate change. That requires work between natural and social scientists to generate a truly interdisciplinary imagination with respect to climate change, while continuing to respect what physical science and each of the other contributing disciplines (social as well as natural) have to offer by themselves concerning different aspects.

Moves in this direction are already taking place in relation to climate change. For example, Integrated Assessment Models (IAMs) of climate change impacts have gained ground since the start of this century. Such assessments apply the judgment of scientists from diverse disciplines to existing knowledge to provide scientifically credible answers to policy-relevant questions (Leemans 2008). The IPCC Working Group III third assessment report of 2003 on 'Mitigation' had this to say of its use of IAMs:

> Researchers have also assessed the costs of climate protection by considering both the economic and biophysical systems, and the interactions between them. IAMs do this by combining key elements of biophysical and economic systems into one integrated system. They provide convenient frameworks to combine knowledge from a wide range of disciplines. These models strip down the laws of nature and human behaviour to their essentials to depict how increased GHGs [Greenhouse Gases] in the atmosphere affect temperature, and how temperature change causes quantifiable economic losses. The models also contain enough detail about the drivers of energy use and energy–economy interactions to determine the economic costs of different constraints on CO_2 emissions (see, e.g., Shogren and Toman 2000; IPCC 2003: 490).

The fourth IPCC Assessment was published in 2007, and includes a synthesis report of the diverse natural and social scientific inputs (IPCC 2007) and one has recently been published as this book goes to press to accompany the fifth IPCC Assessment of 2014. The scope of the 2014 Synthesis Report is defined by the IPCC as follows:

> The Synthesis Report (SYR) synthesizes and integrates material contained within IPCC Assessment Reports and Special Reports. The SYR should be based exclusively on material contained in the three Working Group Reports and Special Reports produced during the 5th or previous Assessment Cycles. It should be written in a "non-technical style suitable for policymakers and address a broad range of policy-relevant, but policy-neutral questions".... The SYR should be largely self-contained, but guide readers to the underlying material if they wish to look further. The primary audience for the SYR will be policymakers, in particular from governments, advisors to policymakers, and experts. However, it is recognized that others will also make use of the report. (IPCC 2014).

We stress once more that the scientific disciplines (natural and social) have individually made enormous contributions to the depth of understanding about climate change. That has been, and still is, hugely necessary. As we stated in Chap. 4, we are not, anti-discipline. We respect the different kinds of knowledge produced within the natural and social scientific disciplines. We do not, therefore, advocate destroying existing knowledge boundaries (as advocates of *transdisciplinarity* might prefer).

That said, we further contend that these disciplinary contributions, impressive as they are, are not sufficient for a complete, integrated knowledge that benefits the climate change debate. In this sense, disciplinary boundaries are limiting, just as norms of attention impose limiting boundaries on lived experiences. Instead, we promote *interdisciplinarity* that welcomes movement across the boundaries in order to enrich knowledge without destroying them. In other words, interdisciplinary approaches provide a route for creatively crossing boundaries and developing a social imagination with respect to climate change.

We have also argued elsewhere for an *expanded* form of interdisciplinarity where not only the sciences engage with each other, and not only people with diverse lived experiences engage with each other, but also scientists from diverse disciplines engage with lived experiences (Abbott and Wilson 2012; Wilson 2012). The reason for doing so is still the same—to obtain a broader, deeper and integrated knowledge of climate change. Thus, for example, while lived experience by itself is a restricted knowledge, it simultaneously offers an expanded insight to that offered by each of the sciences, and vice versa. In other words, for us the central problem is not individual human behaviour or scientific framing. Rather, it is the ability or inability to cross knowledge boundaries between diverse lived experiences, between the scientific disciplines and between lived experiences and the scientific disciplines.

The challenges of enacting science-citizen engagement of this nature are, however, demanding as we saw in Chap. 7 where we covered some of the related literature around this topic. The concepts by which scientific disciplines are disciplined are codified, written down and explained (even when contested) for all to see. The norms of attention that boundary lived experiences are, on the contrary, diffuse, tacit and slippery in the sense of forever being adapted to the particular conversation taking place. These are two very different approaches to knowledge, which we might usefully summarise as scientific and common sense approaches, the latter pertaining to what makes sense to us in a human context (Midgley 2014: 3). Fundamentally different they may be, but they are not necessarily incongruent (Ibid: 1–3) and may even be complementary.

8.4 How Might Expansion of Climate Change Knowledge Be Enabled and Who Might Enable It?

Mills' purpose in 1959 was to promote the idea of a sociological imagination among the social sciences and their practitioners, an imagination that he felt was becoming lost among the constituent disciplines, including within sociology itself. In espousing his mission, he was concerned substantively with academia. He appears silent on the question of how what we have re-named a social imagination develops, or is developed among the general population. In that it is implied in his book, it is the job of social science and social scientists of academia to cultivate it in populations, hence the adjective 'sociological'. Indeed a significant part of the work is an attack on the increasing instrumentalism and particularity of the social science of his day which, in his view, was failing in the supreme task of cultivating this imagination.

An heir to Mills' notion of sociological imagination within academia today is creating interdisciplinary knowledge through multidisciplinary research teams that work on a pressing social issue, for example climate change (see also Sect. 8.3). The shared experience of working together on such an issue is an incentive for discussion and debate around different ideas. It is a counter to the maxim of Brewer (1999: 328), 'The world has problems, but universities have departments'. In this, interdisciplinarity is often encouraged and incentivised by priorities of research funding councils.

There is also some movement towards interdisciplinary teaching in higher education, although in practice this often means 'multi-disciplinary' where students attend classes and seminars led by different disciplinary specialists. Distance learning seems to provide more fertile opportunity for genuine interdisciplinary teaching, where resources are jointly produced by teams. One example of a pan-European interdisciplinary teaching endeavour was 'The Lived Experience of Climate Change' (LECHe) Project that brought together six European Union countries and a large number of scientific disciplines under one umbrella to create Masters-level curriculum. The e-learning modules of the curriculum have since become Open Educational Resources. See Chap. 6, Box 6.2 for further elaboration of this project.

Being of his day Mills was something of an elitist, and a significant part of his book involves shooting down what he considered to be the banalities of other social scientists. Meanwhile, his references to teaching and the role of teaching in higher education are somewhat brief and to the point. Thus he declares at different points:

> To some extent students are a captive audience; and to some extent they are dependent on their teacher who is something of a model to them. His foremost job is to reveal to them as fully as he can just how a supposedly self-disciplined mind works. The art of teaching is in considerable part the art of thinking out loud but intelligibly. (Mills 1959/2000: 79).

> To write for them [students] is to write for a rather specific kind of middle-class public. Author and public, teacher and student, have had similar social experience. They share where they have come from, where they are going, and what might stand in their way. (87)

> In the social sciences, many students agree that 'we ought to get them together' (119).

All of this is fine, but to take these three citations in turn:

The first quotation prompts the question: what kind of conversations might we expect from people in everyday life who share their lived experiences of climate change? Unless they are also teachers in academia, one can hardly expect such people—the vast majority—to have a honed art of thinking out loud but intelligibly with one another. We deal more with this below.

The second quotation has shades of Habermas' notion of 'background consensus' that lies behind all narrative exchange, which we discussed in Sects. 8.1 and 8.2. Today, we might question, however, whether the students and their teachers have indeed all had a similar, middle-class social experience that enables such a consensus. Certainly at least in the affluent countries of the world, higher education has made serious attempts to expand and broaden social access to universities.

Also this second quotation concerns lecturers writing books that are intended to be read by students. Especially in edited volumes ('Readers') where individual chapters are written by different authors, it is possible to present a diverse set of disciplinary inputs on a topic such as climate change, and also include chapters that are written by practitioners. For example, we, the authors of this book, were involved in one such edited volume that included geographers, physicists, environmental scientists, ecologists, sociologists, economists, engineers and practitioners (Wilson et al. 2009). Titled 'Environment, development and sustainability: perspectives and cases from around the world', the book declared boldly in its introduction that diversity should not be considered a problem but productively 'The many perspectives [on the topic] are a resource for us to learn from, gain knowledge and thereby act appropriately'. (Ibid: 8).

The third quotation suggests a degree of pull from students for interdisciplinary curriculum. However, Mills means interdisciplinarity across the social sciences alone. Indeed, elsewhere in the book he is clear that he is specifically interested in students of sociology as the primary discipline with potential for inculcating the 'sociological imagination'.

The moves within higher education towards interdisciplinary curriculum are significant, and climate change is a prime example where such curriculum has the potential to bear fruit. We have noted that significant bodies of both lecturers and students are in favour of it, and sometimes there are material incentives from external funding bodies, as in the LECHe example above. Institutionalisation, however, is a messy process. Individual disciplines and their proponents still carry a great deal of authority, and there is a continual counter-pull of retrenchment to within disciplines, especially when finances are tight, and interdisciplinarity becomes institutionally a kind of expendable experiment.

Others, such as Norgaard (2011) have taken the concept of sociological imagination and applied it to people in their everyday lives. So did Paulo Freire (1972) in his *Pedagogy Of The Oppressed* (see Sect. 8.2). And so have we, and, like Norgaard, we are in the process of doing so in relation to climate change. Implied by these other scholars, and made explicit by us, is to turn the concept into a 'social imagination' in order to loosen the specific ties with the discipline of sociology, which is, after all, a discipline with its own set of concepts like the others.

The challenges for people, as part of everyday life, to engage in deep and broad conversations about lived experiences are much greater, however, than in formal higher education. Messy and slow-moving as developments in interdisciplinary research and teaching might be, there is still some pull and push for them and espoused institutional support. In contrast, in everyday life, boundary crossing conversations will happen occasionally between very good friends (Sect. 8.1), they might happen arising out of shared practical experiences between people who do not know each other very well, but it is highly unlikely that there will be a sustained critical mass over time to continue and reinforce them. To give two examples from the personal lived experience of one of us:

1. I have two friends whom I have known for many years and see about once a month. They are partners and have higher educational qualifications in landscape design. Whenever I go hiking with them, they appear to dislike intensely the wind turbines that keep popping up on hitherto farming land and rough pasture in the hilly, semi-rural area of Northern England where they live, and they are very critical of the planning laws that allow them. I, on the other hand, cannot become upset about wind turbines and equate them in an ill-thought-out way with at least a process of carbon reduction and hence climate change mitigation. We have had one or two slightly heated conversations on the topic, the last one ending with me being asked if I would be happy to see the turbines stretched out end-to-end on the skyline. That was a few months ago but generally we have just agreed, as friends in so many other ways, to differ, and the subject rarely surfaces.
2. I am a fanatical hiker. With my partner, I have twice traversed the Pyrenees range of mountains that form the border between Spain and France from the Atlantic Ocean to the Mediterranean Sea. On our treks, we meet other people sharing the same experience, especially in the basic mountain lodgings that we use for night accommodation and communal meals. We converse with them—there is little else to do—and meet a wide variety of people that way, people who share intensely one thing together—the mountains of the Pyrenees. One evening in September 2014 we sat down for dinner with two Germans about 30 years younger than us and two French about 10 years younger. Part of our conversation was about climate change, triggered by the discovery that one of the Germans works on the design of renewable energy schemes. Of course, I was highly interested but so too were the others around the table. We exchanged views and information freely. There was, however, no follow-up. The conversation changed eventually to another topic and the next morning we all shook hands warmly and went our separate ways as if it had never happened.

It is of course ridiculous to think that developing a social imagination about climate change may happen by itself, spontaneously through chance conversations. Just as in interdisciplinary education there has to be a milieu of people wanting actively to engage, a congenial place to do so, and the opportunity for repeated, reinforcing engagement. Triggers that are grounded in real life are also important. This might be for example, the communal experience of a severe weather event as we saw in relation to UK flood victims in Chap. 1, but there are many different kinds of 'happening' that trigger conversation.

In other words, there has to be a reason for engaging on the topic. Even then, there is no guarantee that engagement will lead directly towards developing a social imagination. In all probability it would not, but it stands a chance of being one step in the process and an important step, because without it, outside of education, no engagement at all will almost certainly mean no social imagination.

Thus, from casual conversation we have to shift our horizons towards purposeful dialogue. Regan (2007: Kindle location 2956) cites Otto Scharmer of the Massachussetts Institute of Technology Global Leadership Initiative who characterises the goals of purposeful engagement in dialogue with respect to climate change:

- To broaden the perspectives of those whose knowledge or expertise is confined to one or another area of specialisation
- To provide opportunities for non-specialists to gain knowledge to which they would otherwise have access
- To deepen each participant's understanding of how a particular problem or its proposed solutions may affect them and others
- To expand the range of imagined possibilities beyond one's own experience or prejudice.

Taken together, these goals provide another way in which we might view acquiring a social imagination.

There are many examples of purposeful engagement and the two edited volumes that we analysed in Chap. 7 provide a good selection in relation to climate change (Moser and Dilling 2007a; Whitmarsh et al. 2011), albeit restricted to a few affluent countries. Below we provide a few others. Ours do not necessarily concern engagement on climate change directly, but they are all on aspects of environmental change and related matters. Three of the examples are broadly positive, while two present challenges. We have chosen them to illustrate the diversity of purposeful engagement, the challenges as well as the potential, and also to provide some general pointers that we may carry forward to Chaps. 9 and 10 which concern lived experience, public action and the making of policy on climate change':

Example 1 Masters-level curriculum in some universities that is aimed at management (including environmental management and management of international development) centres around the concept of the reflective practitioner. That is, the curriculum is aimed to a large extent at existing practitioners who are required to reflect on their theoretical and technical teaching in terms of their own experience of managing and vice versa. We, the authors of this book, once conducted research where we analysed the Masters dissertations of students on a Development Management Programme at the United Kingdom Open University. The research concerned the extent to which the students as development managers engaged in reflective practice in managing interventions that involved engagement with diverse stakeholders. We found that many of them did reflect deeply, but also that the programme developed a community of practice among the students who were drawn from African countries as well as the UK (Abbott et al. 2007).

In terms of this book, we may conclude from Example 1 that higher education has a role to play beyond interdisciplinary teaching, to provide a space for linking directly theoretical knowledge (in this case of management) with the experiences of practitioners, and to encourage critical reflection on each. Of course, this only involves a small segment of the student population, and an even smaller segment of the general population, but that does not lessen its usefulness.

Example 2 Geography field trips to distant places and cultures constitute another potential space within higher education for purposeful engagement with diversity. One of the authors of this book, however, has critiqued these trips and their failure to stimulate any kind of deep political and critical inquiry among participants. Drawing on personal experience of having organised field trips from her UK University to The Gambia, a poor, small country in Western Africa that was once a British colony, she concludes:

> A distinct feature of geography fieldwork is the 'whiteness' of geography rooted in an imperial history. Whiteness reinforces racism, stereotypes, and racist patterns of thought through the empty spaces that result from silence. However, 'whiteness' and its political implications remain unexplored in related literature. Are we then really giving students a valuable experience if we cannot ourselves explore the deeper meanings of fieldwork? One dramatic response to the 'awkward' questions posed by such fieldwork is to abandon it. In a better response, we can work towards a political analysis of long-haul field work, uncover accounts, discover new narratives, and do better and more critical research about geography's historical role as the purveyor of boundaries on spaces of power and hierarchies, and its ramifications onto present day fieldwork. (Abbott 2006)

Thus, even when the object is to expose to and explore difference through practical fieldwork, which after all is what geography as a discipline is predicated on, it is possible, and not unusual, for the space provided to ignore any deep questions and, far from developing a social imagination, to obscure it.

Example 3 Johnson and Wilson (2009: 65–86) researched 'practitioner-to-practitioner' partnerships between United Kingdom environmental health officers and their counterparts in Ugandan municipalities. These partnerships involved municipal engineers working on joint projects, who 'shared the same language' yet had a gulf of cultural and material differences between them. What they shared as engineers provided the elements of 'background consensus' for being able to bounce ideas off each other, while their differences were sources of wider learning.

Other factors that enabled relatively free and open discussion included the shared endeavour of working on joint problem-solving which facilitated trust between the practitioners. The UK practitioners further pointed out the value of having to return to basic principles of engineering when working in the much poorer municipalities of Uganda and discard the work manual that they would use domestically. They were also forced to question why they were doing this work, which led them to reflect on the fundamental values of civil engineering as they developed in nineteenth century Britain, and which led to the grand public health developments of that era, especially around water and sanitation.

This example again throws into sharp relief the need for commonality to be able to engage at all, but also the need for sharp differences to lead to social learning. Further, it illustrates the downside of 'projects', for when the engineers separated for several months in between visits either to Uganda or the United Kingdom, frustration often set in as communication and engagement fell away. On top of that was the difficulty of sustaining projects once funding ended.

Example 4 The Southern-African NGO 'Environment Africa' is based in Harare, the capital city of Zimbabwe. Despite difficult political and economic circumstances, the NGO has been active in setting up Environmental Action Groups (EAGs) in towns, which typically comprise the local municipality, relevant private sector businesses, and local non-governmental groups. Their activities are diverse, depending on the local context, but the important point for us is that, whatever the substantive interest of each, they provide an important space for purposeful discussion between diverse stakeholders (Ibid. 2009: 44–64). Environment Africa has also established 'clusters' between firms in Harare, where rather than the usual reason for such clusters which is to facilitate economic development, they have included a social, environmental mission with respect to maintaining a clean water supply in the city (Hewat and Banda 2009). Again they provide a space for purposeful engagement between actors.

In facilitating EAGs and clusters, Environment Africa appears to be a classic 'boundary organisation' that is able to facilitate engagement across diverse stakeholders, and bring together joint interests (and accommodate different interests) that help sustain them. Without its driving force, and especially that of its 'boundary agent' founder, it is almost inconceivable that any of this could have happened (Hewat and Banda 2009). See Chap. 10 for further discussion of Environment Africa as a boundary organisation.

Example 5 Our final example concerns the possibilities for purposeful engagement about climate change through social media. We illustrated the possibility in Chap. 6 when we relayed the rich discussion about reducing one's personal carbon footprint among Masters students from different European Union countries. This positive example worked because it took place in a structured educational setting and was firmly facilitated. In the public social media, however, interaction on matters such as climate change can be disappointing. One issue concerns our insecurity in publicising viewpoints in front of people we have never met. Another is the opposite, to use the forum boldly and simply to state prejudices and positions with no intention of engagement. Added to this, the very structure of much public social media does not allow for anything but surface engagement with a topic such as climate change. They are designed for 'bite-sized' contributions, that is, very small inputs at a time. Finally, even where significant space is allowed to engage more deeply, the opportunity is often not taken. We do not yet have the skills to express ourselves deeply and effectively without the props of body language and physical proximity that face-to-face engagements bring.

One of the conclusions concerning attempts to develop virtual communities of practice among professionals in a particular domain across the world is that social media may be used to support such communities, but it cannot create or replace them. Face-to-face meetings are needed to establish trust and relationships among participants (van der Velden, cited in Johnson and Wilson 2009: 124). Again, we discuss further the challenges of social media with regard to engagement on climate change in Chap. 10.

None of the above examples make any claim to have 'delivered' a social imagination. That in no way detracts from them, because in their different ways they show what 'might be' and sometimes what 'should not be'. We end this chapter by drawing from them some pointers for the conditions that enable good, purposeful engagement:

- Structured spaces, including educational spaces, for discussion that are also facilitated by an 'independent' person and which include a grounded trigger topic for discussion (for example, comparison of personal carbon footprints).
- A boundary organisation containing the skills and motivation for transboundary work among diverse participants. While Environment Africa sets up EAGs and industrial clusters with a social purpose in a poor, politically divided country, is there not a possibility for replicating elsewhere, including in affluent countries?
- Trust among participants to share their views honestly. Trust may be developed through shared experience of action and through a background consensus of shared language and meanings. However, while the commonality of background consensus is important in this regard, difference between participants is simultaneously the driver of social learning and new knowledge.
- Repeated reinforcement is needed but it is difficult to sustain where participants have limited chances to meet face-to-face and where the projects that may have brought people together in the first place have a boundaried shelf life.
- Not depending on virtual social media. They may support face-to-face engagement but cannot at present replace it.
- Not shying away from unsettling, uncomfortable issues. Great opportunities may be lost if not seized and the engagement becomes frustrating or worse, non-existent.
- Remembering that, away from the actual engagements, people reflect privately and that this too is an important space.

References

Abbott, D. (2006). Disrupting the 'whiteness' of field work in geography. *The Singapore Journal of Tropical Geography, 27*, 326–341.

Abbott, D., & Wilson, G. (2012). The lived experience of climate change: Complementing the natural and social sciences for knowledge, policy and action. *International Journal of Climate Change: Impacts and Responses, 3*(4), 99–114.

Abbott, D., Brown, S., & Wilson, G. (2007). Development as reflective practice. *Journal of International Development, 19*, 187–203.

Borda Rodriguez, A. (2008). *Knowledge for Development?Reflections from Consultants and Advisors in Bolivia*. Ph.D. thesis, The Open University, Milton Keynes.

Brewer, G. (1999). The challenges of interdisciplinarity. *Policy Sciences, 32*, 327–337.

Chataway, J. (2005). Introduction: Is it possible to create pro-poor agriculture-related biotechnology? *Journal of International Development, 17*(5), 597–610.

Chipika, S., & Wilson, G. (2006). S. Enabling technological learning among light engineering SMEs in Zimbabwe. *Technovation, 26*(8), 969–979.
Edgar, A. (2006). *Habermas: The Key Concepts*. London: Routledge.
European Commission. (2010). Agenda for new skills and jobs. http://ec.europa.eu/social/main.jsp?catId=958. Accessed 24 February 2015.
European Commission. (2011). A Roadmap for moving to a competitive low carbon economy in 2050. http://eur-lex.europa.eu/resource.html?uri=cellar:5db26ecc-ba4e-4de2-ae08-dba649109d18.0002.03/DOC_1&format=PDF. Accessed 24 February 2015.
Fischer, F. (2003). *Reframing public policy: Discursive politics and deliberative practices*. Oxford: Oxford University Press.
Forbes, N., & Wield, D. (2002). *From followers to leaders: Managing technology and innovation*. London: Routledge.
Freire, P. (1972). *The Pedagogy of the Oppressed*. London: Penguin Books.
Habermas, J. (1984). *The theory of communicative action (Volume 1): reason and the rationalisation of society*. Cambridge: Polity Press.
Habermas, J. (1987a). *Knowledge and human interests*. Cambridge: Polity Press.
Habermas, J. (1987b). *The theory of communicative action (Volume 2). Lifeworld and system: A critique of functionalist reason*. Cambridge: Polity Press.
Habermas, Jürgen. (1990). *Moral Consciousness and Communicative Action*. Cambridge: Polity Press.
Hewat, C., & Banda, B. (2009). Innovative partnerships for sustainable development in Harare, Zimbabwe. In G. Wilson, P. Furniss & R. Kimbowa (Eds.), *Environment, development and sustainability: Perspectives and cases from around the world* (Chapter 22). Oxford: Oxford University Press.
Hulme, M. (2009). *Why we disagree about climate change*. Cambridge: Cambridge University Press.
Hulme, M. (2013). *Exploring climate change in science and society: An anthology of Mike Hulme's writings, interviews and speeches*. London: Earthscan/Routledge.
IPCC. (2003). *Climate Change 2014: Mitigation of Climate Change. Contribution of Working Group III to the Fifth Assessment Report of the Intergovernmental Panel on Climate Change*. Cambridge, United Kingdom: Cambridge University Press.
IPCC. (2007). *Climate Change 2007: Synthesis Report*. Cambridge, United Kingdom: Cambridge University Press.
IPCC. (2014). *Synthersis Report AR5*. http://www.ipcc-syr.nl/. Accessed 24 February 2015.
Johnson, H., & Wilson, G. (2009). *Learning for development*. London and New York: Zed Books.
Leemans, R. (2008). Personal experiences with the governance of the policy-relevant IPCC and millennium ecosystem assessments. *Global Environmental Change, 18*, 12–17.
Machin, A. (2013). *Negotiating climate change: Radical democracy and the illusion of consensus*. London: Zed Books.
Midgley, M. (2014). *Are you an illusion?*. Durham: Acumen.
Mills, C. W. (1959/2000). The sociological imagination. Oxford: Oxford University Press.
Mohan, G., & Wilson, G. (2005). The antagonistic relevance of development studies. *Progress in Development Studies, 5*(4), 261–278.
Moser, S., & Dilling, L. (Eds.). (2007a). *Creating a climate for change: Communicating climate change and facilitating social change*. Cambridge: Cambridge University Press.
Moser, S., & Dilling, L. (2007a). Toward the social tipping point: Creating a climate for change. In S. Moser & L. Dilling (Eds.), *Creating a climate for change: Communicating climate change and facilitating social change* (Chapter 32). Cambridge: Cambridge University Press.
Norgaard, K. (2011). *Living in denial: Climate change, emotions and everyday life*. Cambridge, Massachusetts: The MIT Press.
Picketty, T. (2014). *Capital in the twenty-first century*. (A. Goldhammer, Trans.). Cambridge Massachussetts and London: The Belknap Press of Harvard University Press. (Originally published as: Picketty, T. 20113. *Le capital au XXI siècle*. Paris: Editions du Seuil).

Regan, K. (2007). A role for dialogue in communication about climate change. In S. Moser & L. Dilling (Eds.), *Creating a climate for change: Communicating climate change and facilitating social change* (Chapter 13). Cambridge: Cambridge University Press.

Shogren, J, & Toman, M. (2000). P. Portney & R. Stavins (Eds.), *Climate Change. Public Policies for Environmental Protection* (2nd ed.). Washington, DC: Resources for the Future Press.

Voltaire, F. (1759/2005). *Candide, or optimism.* With Introduction by M. Wood, (T. Cuffe, Trans.). London and New York: Penguin Classics.

Whitmarsh, L., Lorenzoni, I., & O'Neill, S. (Eds.). (2011). *Engaging the public with climate change: Behaviour change and communication.* London: Earthscan.

Wilson, G., Furniss, P., & Kimbowa, R. (Eds.). (2009). *Environment, development and sustainability: Perspectives and cases from around the world.* Oxford: Oxford University Press.

Wilson, G. (2012). The lived experience of climate change: Expanding the knowledge base through collaborative Master's curriculum in the European Union. *International Journal of Innovation and Sustainable Development, 6*(1), 43–52.

Chapter 9
Rationalist and Public Action Theories of Knowledge in Climate Change Debates

Abstract Whatever our uncertainty about climate change knowledge, what matters is how we respond to this and what action we take. Devising appropriate action for policy for climate change generates much debate and disagreement. This chapter discusses differing approaches to policy making for climate change, beginning with dominant rationalist, mostly linear, approaches. It suggests that both scientific positive and normative assertions that are embedded in a rationalist approach are also reflected and reinforced through discussions, often heated, on action and policy making. However, prescriptive linear policy 'from above', surfacing from a 'nanny state', is often resented by the public. Policy also needs public support which determines how far governments can enforce it when there are several tensions around finding common ground, conflicts over resource priorities and ethical and moral issues of social justice and equity. Also, individual rationalisation based on personal perceptions and lived experience can be more powerful, overriding, challenging and cancelling rationalist contention. Altogether, such considerations generate several problems for a linear rationalist approach. For these reasons, a preferred alternative is a non-linear public action approach which arguably has a better, wider fit with how policy is actually made. In this approach, policy at all scales is a perpetual social process of being made and remade, rather than shaped through an expert-led prescription. It recognises that robust policy and intervention is constructed from the engagement of both scientific knowledge and lived experiential knowledge. There are, however, real issues of power at play in such engagement, concerning whose knowledge counts, and the political use of contested knowledge to cancel evidence claims. Nevertheless, despite the challenges, difference and diversity are sources of social learning and new knowledge. As Chap. 8 has argued, we require a productive transboundary engagement and social imagination to expand knowledge boundaries. We don't learn by being the same.

9.1 Introduction

Whatever our uncertainty about climate change knowledge, the underlying issue concerns how to respond to what we know and what action we can take within

D. Abbott and G. Wilson, *The Lived Experience of Climate Change*,
DOI 10.1007/978-3-319-17945-2_9

the context of other overarching issues such as resource availability. Questions on action and response, and devising appropriate policy for climate change have generated much debate and disagreement over recent years.

First, note that there are many definitions of policy. Here are two, which encapsulate it in general terms: (i) a course or principle of action adopted or proposed by an organisation [for example, a private company, NGO or governmental department] or individual (Oxford Dictionary); (ii) a deliberate system of principles to guide decisions and achieve rational outcomes (Wikipedia). When, however, we refer to policy in this book we mean policy of government (at national and local levels, and of individual departments) or policy arising from intergovernmental agreement.

This chapter starts in Sect. 9.2 by illustrating the tortuous and conflictual process of international climate change negotiation over the Kyoto Protocol. As Sect. 9.3 argues, however, conflicts over climate change policy also exist at national and local scales, all of which are inter-connected. Behind these policy conflicts are attempts to reach accommodation between positive scientific and normative social and political assertions, and the role that the media play in communicating scientific information (Sect. 9.4).

Conflicts at every scale do not fit easily with the traditional, rationalist approach to analysing policy making (Sect. 9.5). This approach considers policy to be a prescriptive outcome of a linear process that is informed by scientific, technical expertise and 'value-free' facts. All policy prescriptions, however, draw a certain amount of disquiet amongst the public. Prescriptive linear policy 'from above', for instance, is often resented by the public who may view it as surfacing from a 'nanny state'. Moreover, if the actions that emanate from policy are against the interests of certain groups of people—for example, those who are directly affected adversely by climate change mitigation attempts—the policy will be actively opposed. Thus, policy also needs public support. At times the extent to which governments can legislate depends on differing short- and long-term public interests, tensions between finding common ground, conflicts over resource priorities and ethical and moral issues of social justice and equity amongst other things. At the end of the day, however, individual rationalisation based on personal perceptions and lived experience can be more powerful, overriding, challenging and cancelling rationalist contention.

For these reasons, this book prefers a non-linear public action approach as having a better, wider fit with how policy is made in practice (Sect. 9.6). In this approach, policy is a perpetual social process of being made and remade, rather than shaped through an expert-led prescription. It recognises that robust policy and intervention are constructed from the engagement of both scientific knowledge and lived experiential knowledge. There are, however, real issues of power at play in such engagement, concerning whose knowledge counts and the political use of contested knowledge to cancel evidence claims (Sect. 9.7). Further issues of power underlie attempts at public engagement with science, including resistance to internalising the science of climate change, but Sect. 9.8 also provides an example of positive public action through demonstration of local innovations. The latter illustrates that, despite the challenges, difference and diversity are sources of social learning and new knowledge. We do not learn from being the same.

9.2 Towards a Targeted Climate Change Policy

Climate change as a major global issue became mainstreamed internationally and politically in the late 1980s (Chap. 3), although public concern in a few countries was growing before then. The broad concern that anthropogenic release of gases into the atmosphere may cause public harm, however, emerged a decade earlier. This was the realisation that chlorofluorocarbons (CFCs), used typically in aerosol sprays and cleaning fluids, and as a refrigeration gas, depleted stratospheric ozone, a gas that is important in filtering ultraviolet rays from the sun. The 'ozone hole' as it was called was allowing this radiation through with adverse effects on ecosystems, and human health and well-being—of which skin cancer was the most publicised.

Action to ban CFCs took a while, however, and it was only in 1987 that The Montreal Protocol on international cooperation for the protection of the ozone layer was negotiated by 24 countries and those who formed the European Economic Community (the forerunner of the European Union), coming into force in 1989. The Protocol prescribed a phase-out of CFCs and other man-made 'Ozone Depleting Chemicals' (ODCs), leading to an agenda for strict regulatory controls by the countries involved. Whilst there was a critique of action (or even lack of action), the result nevertheless suggests that by the late 1980s and 1990s, industrialised nations had mostly eradicated these harmful chemicals. It was for these reasons that Kofi Annan, in the United Nations (UN) Millennium Assessment Report of the Secretary-General (2000: 56) claimed: 'Perhaps the single most successful international environmental agreement to date has been the Montreal Protocol, in which states accepted the need to phase out the use of ozone-depleting substances'.

Woodhouse (1992: 98–99) makes an interesting point about the apparent success of the Montreal Protocol where his description of the process in reaching agreement resonates with the subsequent international wrangling over climate change, but also contrasts with it by achieving a positive outcome. The United States, somewhat of a laggard in international climate change negotiations, was actually a key driver of the Protocol. This was because it already had developed substitutes for CFCs and thus had a simple technical solution to the problem. This solution, moreover, did not involve big lifestyle changes or threats to the national energy economy that is at the centre of US resistance to binding climate change action. China and India, on the other hand, were more reticent, arguing that banning CFCs would effectively interfere with their countries' development and that they were being asked to clear up the mess created by rich countries. They only came on board on the understanding that the substitute technologies would be made available to developing countries.

Turning to climate change, and especially the international dimension, in 1988 the UN General Assembly (1988) officially fortified the 'need for additional research and scientific studies into all sources and causes' of the phenomenon. It recognised that 'climate change is a common concern of mankind, since climate is an essential condition which sustains life on earth' and determined that 'necessary and timely action should be taken to deal with climate change within a global framework'.

The pressing need to further scientific knowledge for the 'common concern for mankind' brought Governments of participating countries together to create the Intergovernmental Panel on Climate Change (IPCC 2014) in 1988 (see also Chaps. 3 and 6). According to the IPCC website:

> It was set up by the World Meteorological Organization (WMO) and the United Nations Environment Program (UNEP) to assess and synthesise available scientific information on all aspects of climate change and its impacts, with a view to formulating realistic response strategies. The initial task for the IPCC as outlined in UN General Assembly Resolution 43/53 of 6 December 1988 was to prepare a comprehensive review and recommendations with respect to the state of knowledge of the science of climate change; the social and economic impact of climate change, and possible response strategies and elements for inclusion in a possible future international convention on climate (IPCC history).

At the same time, serious effort was being made by various countries for taking action in response to climate change. Many international conferences attended by national, senior, prominent representatives of country governments came together to negotiate an international treaty that became known as the United Nations Framework Convention on Climate Change (UNFCCC 2014), which was signed in June 1992. By 1994 it had established policy-making guidance to stabilise greenhouse gas concentrations in the atmosphere with an important recognition for 'common but differentiated responsibility'. Thus Article 3 of the UNFCCC states that 'parties should protect the climate system for the benefit of future and present generations of human kind on the basis of equity and in accordance with their common but differentiated responsibility and respective capabilities. Accordingly, developed countries should take the lead in combating climate change and the adverse effects thereof' (McManus 2009: 2). Specifically, the UNFCCC requested this group of countries to stabilise its greenhouse gas emissions to 1990 levels by the year 2000. This proposition clearly demarcated responsibility between the poor developing and the rich developed countries—a bone of contention for both for many years to come.

Undoubtedly, the UNFCCC treaty had a high starting ambition in stabilising greenhouse gas emissions, but it had no mandate to set binding limits on emissions from individual countries. Not surprisingly, for various reasons including political commitment and capacity, all Governments were unable (or unwilling) to meet the general targets that it set, and extensive debates took place on what stronger measures could be established and how they could be executed.

In the final analysis, however, UNFCCC was conceived, and remains, a framework for negotiation about stabilising greenhouse gas emissions. In this vein it organises an annual 'Conference of the Parties' (COP) where the parties are those states that have ratified the general treaty—196 in 2014. The concern about lack of legally binding targets was raised at the first COP in 1995 and culminated in the signing of the Kyoto Protocol in December 1997 which sets emission targets for developed, industrialised countries that are legally binding in international law. States bargained hard with raised voices over emission limits and the demarcation of level of responsibility between developed and developing countries. There was particular discontent emanating from the United States and the rapidly industrialising and hugely populated developing world countries, especially India and China. While the European Union tried to broker agreement, the US hostility toward the Protocol under the Bush administration led to a breakdown in 2000 and it still has not signed up as we write. Nevertheless, over a number of years the Protocol did obtain the signature of most member states (192 in early 2015), including 37 industrialised countries that agreed to binding commitments and which are listed in its Annex I.

Disputes have continued over the efficacy of the Protocol at successive COPs. The 2009 COP in Copenhagen was particularly damaging as expectations had been raised beforehand that all countries would set aside their differences to work substantively towards a more robust binding climate change agreement for the common good of human beings everywhere. This was not achieved and in a disappointing finale to the conference, emission reduction, mitigation-action pledges found their way into a Copenhagen Accord as it is known, but remain non-binding.

The first commitment period for signatories to the Kyoto Protocol started in 2008 and ended in 2012. The second commitment period started on 1 January 2013 and is planned to end in 2020, but is seen very much as an interim measure until a suitable and better alternative is established. Already some countries have withdrawn from the second commitment period (joining the United States which never signed up in the first place) and there is doubt that the obligations of those remaining will be met. There is hope that continuing negotiations and amendments to the Protocol will result in improved and more robust greenhouse emission reduction targets by 2030.

In spite of the recognition that climate change is a 'common concern of mankind', individual and national diversity and self-interest has thus complicated our history over mitigating action. The extraordinary and complex story of Kyoto negotiations is glaring evidence of this. Failure to easily reach compromises has been in a broader context of other global pressing issues such as a world financial crisis, the rise in terrorism and new diseases such as Ebola, all of which demand resources. Consequently, climate change has taken somewhat of a back seat between 2009 and 2015.

What is evident, however, from the debates marked by the 'milestones' of various treaties and protocols over action on climate change is that: (i) any action plans are closely related to science and scientific findings, and (ii) that even if we are aware of the harm that our actions (or rather lack of them) may cause for us and future generations, as human beings with individual and national interests, we continue to quarrel over what international policy should be allowed dictate and how far we are prepared to go. Thus, the most world's richest country (the US) is unwilling to reduce huge per capita energy consumption and emission levels, while hugely populated countries such as India and China are equally unwilling as they undergo rapid industrialisation and demand a place within powerful economic and political power play.

9.3 Positive and Normative Assertions of Policy Making

Section 9.2 has examined policy-making and action around climate change in the international context, which is inherently anarchic because it is based on agreement between sovereign states where there is no global-scale coordinating power. No wonder therefore that it is highly contentious. Meanwhile, at national and local scales the degree of contention appears to be no less as climate change mitigation actions affect communities adversely in the present (Chap. 6).

The reasons for this state of affairs across scales are many. For example, while the science of global warming leading to climate change may be clear, at national and local scales the precise nature of impacts is not. Economics, which plays a large role in any policy deliberation, strives to put numbers to impacts and intervention measures. This task, however, requires assumptions in order to calculate the balance between economic costs and benefits of proposals. As noted in Chap. 3, such assumptions can be themselves highly controversial and therefore contested. On top of this there are other important factors—we are also unsure about how our actions will fit in with our cultural values, societal norms and national interests.

Arguments and uncertainties such as these revolve around both positive and normative assertions (after Dessler and Parsons 2010). We are dependent on positive scientific evidence of climate change to lead the way but are also confined within boundaries of politics and societal values which shape decision-making. Positive assertions thus include general observations of events, causal relationships and trends based on the assumption that something exists beyond us which comprises the 'truth' (Chap. 4). Statements that now form the basis of debate on climate change, for instance, revolve around positive assertions arising from scientific observation and causal relationships between global warming and extraordinary events of ocean activity, flooding and unexpected weather. In contrast, normative assertions are often made by those who query the objectivity of these 'truths' and challenge the ethics, morality, cultural norms and ideologies that are critical to climate change questions. Examples include the excessive fossil fuel consumption of richer countries which, in a globally linked climate, will undoubtedly affect local climatic trends in poorer countries. Also the ethics and morality of a culture of overconsumption within a capitalist economic system which looks to profit from natural resources such as forests rather than preserving them.

A central point of contention is that positive assertions about impact are backed by science and scientific knowledge. If the public do not accept this knowledge and fail as a result to legitimise scientific claims, it can break individual and institutional reputations and jeopardise credibility. Normative assertions, on the other hand, are often intangible, emotional and based on drawing the public in through debates on morality, religious values and doing the 'right' thing.

Usually policy debates and decision-making draw on both positive and normative assertions about an issue. For instance, climate change policy arguments incorporate both science-backed positive assertions as well as normative consequential assertions about societal sustainability. There are, of course, many overlaps and grey areas between both assertion types. We cannot easily unravel the positive from the normative in something as critical and emotive as climate change. Yet a pathway has to be found through the complexity and the perplexity of public debate and understanding if we are to arrive at some agreement on action.

Nevertheless, as citizens and stakeholders, we share a level of common interest in engagement with climate change policy making. We do wish to have a sustainable future for our children. We wish to maintain a decent standard of living

Fig. 9.1 Tree-planting in the United Kingdom. A significant number of people plant trees on plots of land that they own or share. Trees have a role in capturing carbon dioxide from the atmosphere through the process of photosynthesis. They also arrest water run-off, which is important after heavy rain, and help prevent soil erosion. In the UK, tree planting is supported by a charity, The Woodland Trust (https://www.woodlandtrust.org.uk/)

and have a national interest in mitigating climate change events such as flooding and droughts, and we are often prepared to take personal action (Fig. 9.1). We also want to save money, both as individuals and as nations by adopting strategies where the benefits outweigh the costs. At the same time, individuals and groups may wish to use the debate to place themselves on an important national agenda. Ambitious national and local government representatives may thus push personal agendas, be critical and judgemental of others and ensure that the benefits of any policy involve only material incentives, for example the United Kingdom Government's partial or full grant-aided house insulation to preserve heat and payment incentive towards solar panels to reduce fuel use.

9.4 The Role of the Media

The role of the media in how they represent climate change is crucial in policy decision-making because they are instrumental in 'translating' everyday stories or the latest scientific analysis on climate change events and findings. All media reports are value-ridden with a purpose of selling newspapers, increasing viewing figures or reflecting government or other interests such as those of large firms. Thus whilst ethically the media (written, visual and sound) should be reporting 'facts', and theoretically these should come across similarly to all members of the public, editorial intervention and values of the organisation mould the facts accordingly. This may not happen intentionally but, after all, it is in media interest to chase stories that attract the public in larger numbers. However, as the lay public relies more on the media than on sourcing scientific reports, the role they play in relaying climate change information is critical.

As the blog 'Talking Climate' (2015) points out, reporting on climate change is, as expected, further coloured by the political position of the organisation (Painter 2011). The blog goes on to suggest that the left-leaning UK newspaper *The Guardian*, for example, considered to be at the centre of climate change and environmental reporting and opinion, rarely supports climate sceptics. In contrast, the right-leaning *US Wall Street Journal* will give far more space to sceptical voices and editorials. The blog further suggests that, whilst it is difficult to establish direct cause and effect, media ideological leanings will no doubt influence public perceptions, leading to a media-generated controversy over climate change and scepticism (Poortinga et al. 2011). In addition to the influence of ideological leaning, there are other factors that may influence public reaction and scepticism, for example, media 'exaggeration' of climate change (Whitmarsh 2011). Also, the media's ability to present a 'balanced' set of views is in question. As analysed in several places, the aggressive style of journalism does not always go hand-in-hand with the uncertainty of climate change resulting in competing views being equally supported in a 'balance-as-bias' (Boykoff and Boykoff 2004; Boykoff 2007; Ward 2008). Evidence for 'balance-as-bias' comes, for example in a 2011 report commissioned by the governing body of the British Broadcasting Corporation (the internationally renowned BBC) to analyse its reporting of three scientific topics—including climate change (Jones 2011). The report's findings suggest that in seeking a 'balanced' coverage of climate change, the BBC inadvertently ended up with giving excessive time to sceptics through a 'balance-as-bias'.

The 'Talking Climate' blog also cites a report from the Oxford Reuters Institute for the Study of Journalism (Painter 2011) which specifies reporting of climate change scepticism in six different countries (UK, US, Brazil, China, India and France). Remarkably, an analysis of selective newspaper content for each of these countries showed that coverage given to sceptical voices was an 'Anglophone' phenomenon, within the English-speaking UK and US rather than in Brazil, China, India and France. The report suggested that a possible reason for this disparity could be found in organised lobbying interests in the US and the UK, active in seeking to influence the media. Another possible reason (as shown in Chap. 7) could also be the existing high level of general public mistrust in these countries of scientists and other professionals Hulme (2013: 252–264).

The role of the media in the public understanding for climate change, and therefore ultimately for policy and action, is critical, but not always fair or balanced.

9.5 The Rationalist Model of Policy Making

To manage the complexity and contestation in policy making that we have seen so far in this chapter, various models of the process have been developed and used, stretching back decades. A dominant one is the rationalist model with its roots in science and the idea that scientific knowledge is more advanced and more systematic than that of the public. Classic literature on rationalist models may be traced back to Lasswell and Kaplan's (1950) influential works where they argue that the objective 'truth' and validity of scientific knowledge will be able to mitigate conflicting ideologies through its value-free metaphysical position and thus be able to arrive at the 'right' decisions which will benefit all.

As Kuruvilla and Dorstewitz (2009: 2) suggest, 'People across society would prefer that public policy-making is 'rational'. Sound reasoning should make for well-informed decisions and successful strategies. However, different perspectives proffer conflicting opinions on what constitutes rationality. A traditional view in public administration is that rationality is a scientific, or technical, mode of reasoning that is employed to achieve political ends or goals—without questioning the morality, or worth, of these ends.' Whilst, therefore, there is not an agreed consensus on what constitutes 'rationality', the association with science, technical expertise and 'value-free facts' form the basis of a rationalist model and the variations (such as bounded and incremental rationalist models) that have developed for policy studies, but will not be taken up here. It is perhaps sufficient to know that these variations revolve around what is known as the linear rationalist model, rejecting or accepting some of its attributes.

Thus, Kuruvilla and Dorstewitz (ibid) continue, 'In policy theory, linear rationality is reflected in the seemingly ubiquitous 'stages' model. The stages model depicts policy-making as moving through distinct steps of policy agenda setting, formulation, decision-making, implementation and evaluation [see also Chap. 11]. In policy practice, linear rational thinking is evident when public preferences and political positions are invoked as inviolate guides for future policy projects'.

Rationalist policy models remain a common favoured method by governments, and in this, the linear, rationalist model is dominant. This is in spite of the inherent problems and much criticism regarding its value base, ability to represent diversity and so on, to which we now turn. As early as 1959, in response to rationalist theorists, Lindblom (1959) for instance argued that policy making was the 'science of muddling through'. He suggested that the rationalist approach was simply based on masses of information and a comparison of past experiences of smaller policy initiatives to frame them into larger actions for the future.

A critical problem of rationalist policy making is, of course, the uncertainty of the scientific knowledge on which it relies. In *Why we disagree about climate change* Hulme (2009: 83) argues that:

> Uncertainty pervades scientific predictions about the future performance of global and regional climates. And uncertainties multiply when considering all the consequences that might follow from such changes in climate. Some of the uncertainties originate from the

> incomplete understanding of how the physical climate system works – the effects of the atmospheric aerosols on clouds, for example, or the role of the deep ocean in altering surface heat exchanges. …other sources of uncertainty emerge from the innate unpredictability of large, complex and chaotic systems such as the global atmosphere and ocean. At best we are able to estimate some probability….A third category of uncertainty originates as a consequence of human beings being a part of the future predicted. Individual and collective choices …into the future are not predictable in any scientific sense…

Altogether, this makes it problematic for those involved in decision-making to think strategically and confidently about how to assess future risks on climate change based on current knowledge. Hulme (ibid) continues that,

> Although scientific predictions about arcane or merely philosophical questions can afford to be nonchalant about uncertainty, the high stakes, and urgent decisions' character of climate change means that policy makers do not have the luxury of nonchalance. They are faced with the problems of making sensible decisions today in the face of uncertain risks of tomorrow…Yet the curious thing about the relationship between climate change science, knowledge and decision making is that it should be 'uncertainty' rather than 'certainty' that is the condition that is perceived to be anomalous.

Nevertheless, there is urgent pressure to act within the boundaries of what we do know and Hulme suggests, 'even when there is not unanimity among all scientific experts', consensus approaches based on 'expert elicitation and/or large consensus-driven scientific assessments are probably the "least worst" way of progressing' (Ibid: 91). Such a consensus, therefore, does not mean that decisions made on the basis of this are the 'truth' or that the risk assessment thus made is assuring. This tension between half-correct predictions and the half-correct action generated through them does not do the public a favour as this leads to false expectations and doubting of science.

Unfortunately, decisions based on linear rationalist (and associated models) in their reliance on scientific and technical expertise obscure normative concerns of the many stakeholders. There has, for instance, been much focus on techno-fixing and the broad term 'geoengineering' which encompasses several sunlight reflection methods and attempts to reverse the process of fossil fuel combustion (Hulme 2014: 7–8). There have been special efforts to absorb developing countries into a 'climate smart' development especially in sectors such as energy, transport, land use (agriculture, forestry, fisheries), industry, waste, water resources, health (Srinivasan et al. 2012: 223–224). As Hulme (2014: 130), argues, however, 'The dangers of climate change are not somehow 'out there', like a danger of alien invasion—external to the way we live and organise ourselves. The dangers of climate change are 'in here' a function of human technologies, social relationships, economic and political systems'. If the 'in here' context is neglected and there is little focus on citizen knowledge, lived experience, underlying vulnerabilities and explicit inclusion of normative concerns, then rationalist top-heavy linear policy alone cannot generate community engagement.

9.6 The Social Process of a Non-linear Public Action Approach and Lived Experience

The politics of climate change which is played out in a public arena including, of course, the internet is extremely contentious. Concern over action is vocalised by individuals across the world, NGOs and other concerned agencies, as well as several public figures including those from private firms, industry, and Government Ministers who often engage with populist politics for self-benefit.

As we have seen above, the inadequacy of linear rationalist models to portray the actual dynamics of climate change policy-making is illustrated by the difficulties associated with the uncertainty of scientific evidence (including that from economics on costs and benefits), the inability of international protocols to enforce action and the neglect of normative concerns. The internet has, of course, opened global spaces for arguments from opposing 'camps'. Criticism of climate change policy comes in many forms but generally falls within two main 'camps'. There are those who question the role of experts (such as those linked to the IPCC), that they are mainly from the richer countries and thus disassociated from local political, economic and cultural contexts of poorer countries. This is especially so as the dominant (natural) sciences are seen as neutral and do not always adopt a multi-disciplinary strand which may be able to offer a differing analysis. Secondly, there are those who do not believe in global warming at all, or those who focus on natural causes alone, denying the contribution of anthropogenic activity, and yet others who argue that projections of emissions and the damage these can cause are over-estimated and will eventually decline even with minimal effort.

The 'public action' approach is the alternative to linear rationalist policy-making models (Drèze and Sen 1989; Wuyts et al. 1992). Its premise is that policy-making is a social process where at every stage knowledge and action are contested and negotiated through the influences of many interested stakeholders. These range from NGOs, citizen groups, the private sector, professional and worker associations and the media in addition to the state. In climate change policy-making, the stakeholders also surely include the scientific community. Policy is then a social process rather than led by an expert- and state-led prescription.

The examples we gave at the start of this chapter are of the Montreal Protocol on ozone depletion and the protracted machinations over the Kyoto Protocol which aim ultimately to reduce greenhouse gas emissions illustrate the public action process. In these examples, public action is on an international scale where there are many states and other kinds of stakeholder that are trying to influence the (always provisional) outcomes. Note also that the policies that result are in themselves provisional, being continually remade.

In analysing these examples of international processes it is easy to note that the policy that eventually emerges on climate change is through public action that involves power play between the usual suspects—powerful and less-powerful states, alliances between states, international NGOs and the elephant in the room, the global energy industries. States do not, however, act in isolation from their domestic contexts, in particular that of their citizens. Even the most authoritarian of states needs to appear for legitimation purposes (Chaps. 5 and 6) to be acting in their interests. What states perceive to be a kind of general will of their citizens will therefore feed into individual state interests. Thus when President George W. Bush Senior refused to sign the Kyoto Protocol in 2001, he declared that he was acting in the interests of United States citizens and would never do anything to jeopardise their lifestyle.

To the extent that a public general will informs a state's interests exists, it does not arise from nowhere. It in fact represents a synthesis of individual lived experiences of a phenomenon such as climate change. Whilst that may be claimed at the international scale as the refusal of George W. Bush to sign the Kyoto shows, at the national scale, however, we see divisions. Rather than a homogenous 'general will', we see citizens acting to influence policy from diverse interests that are informed by their lived experiences.

Whether at international or national scales, public action and policy, therefore, are not simply the process and result of power play among usual suspects. They also illustrate how lived experience is marshalled collectively among citizens both formally and informally and represents a significant influence. An example of the former would be the organised local campaigns against what are considered to be damaging mitigation actions—such as the installation of renewable energy projects (Chaps. 1 and 6). The latter might be represented in rich countries by forms of 'socially organised denial', which is how Norgaard (2011) characterises the lack of engagement with climate change in a Norwegian village and is recounted in Chap. 7. Here, socially organised denial represents a kind of local public will not to engage with the subject.

We are thus moving towards the suggestion that a lived experience lens is useful for analysing the public action policy process with respect to climate change, both nationally and internationally. This immediately explains the existence of many competing voices in addition to those in specialised knowledge domains. In other words, analysing lived experiences within this process completes the shift from a rationalist to a public action approach on the use of knowledge in climate change policy making and intervention. It enables an alternative approach which acknowledges diversity and contestation of knowledge, inherently has to accommodate lived experiential knowledge, and fits better with how policy and intervention is made in actual practice, whether this be at international gatherings under the auspices of the United Nations or at a grounded, local level.

Lived experience also provides insights into how interests are formed by the many broader groups who seek to be heard at the negotiating table and how they understand, interpret and embrace notions of climate change. This is illustrated by the stories of the two African environmental activists—Charlene and Richard—that we introduced in Chap. 1 who acknowledged the significant influence of their childhood experiences of farming and caring for nature in making them who they are today. We suggest extending this observation to other groups, such as business people and climate scientists, who will have been influenced by historical lived experiences. Nobody is immune from their lived experience and, we add, nobody can set it aside in the name of an elusive objectivity.

There are, of course several methodological challenges of representing lived experience (as discussed in Chap. 4). Nevertheless, the concept explains the differing standpoints (see Box 9.1) that inform different perspectives. For instance, as an editor of several narratives, Dankleman (2010) analyses climate change from a gendered perspective.

Box 9.1 Standpoint

How we understand the exact nature of any issue, and how we go about investigating and interpreting it, depends on our own understanding of the world in which we live. This is defined by multiple, overlapping factors including our place in society and where we feel we belong. For instance, race, caste, class, gender and the power relations these involve may shape the mental position from which we view the world as individuals and/or members of various groups to which we belong. In social science terminology, this is known as 'standpoint' and it shapes our understanding of reality based on our place in society. The underlying argument is that our standpoint is socially constructed, defined and shaped through status and power we hold within our overlapping primary (family, kinship) and secondary (work and leisure peer groups, community groups) societal relationships.

Power and the social constructs of equality and inequality heavily influence our standpoint. A 'classic' and evident example is the artificially created racial divide in apartheid South Africa which gave Black and White South Africans very different societal and economic power, perceptions and standpoint within the same country.

Lived experience analysis helps us to understand why and how people behave in the face of local impacts of climate change. It also illustrates how such behaviour leads to a public need and a public response. Sometimes, in their desperation, the behaviour of those at risk is not one that either the individual or Governments would wish but something completely different. For example whole rural communities might abandon their way of life and migrate to the nearest city, generating a new set of problems both at their new and 'home' locations. No Government is likely to welcome this new kind of 'climate refugee' or people forced to move due to environmental events as can be seen in the example of Bangladesh/India in Box 9.2.

Box 9.2 A global 'national security' issue lurks at Bangladesh's border
Source: Lisa Friedman, (ClimateWire), The New York Times (March 23, 2009)

BHOMRA, Bangladesh—A high heavily reinforced barbed wire fence cuts a jagged line through an otherwise empty field of tall grass and tamarind plants here. Climate change didn't bring this fence but it is providing a fresh reason for its existence and ongoing expansion. On this side of the fence rising sea levels caused by climate change are beginning to inundate low-lying Bangladesh. Scientists estimate that by mid-century as many as 15 million people could be displaced. On the other side of the fence India isn't taking any chances. Already alarmed about illegal immigration it is nearing completion of about 2100 miles worth of high-tech fencing along its long and porous border with Bangladesh….

The prospect of international migration is a touchy subject in Bangladesh. But for national security experts it's the most feared global consequence of climate change. As warming temperatures deplete water supplies and alter land use military analysts warn already-vulnerable communities in Asia and Africa could descend into conflicts and even wars as more people clamour for increasingly scarce resources….

Research on how climate change might spark conflict is still in its infancy and it often tends to be thin and speculative…But in Bangladesh where most of the country is less than 20 feet [6 m] above sea level many analysts say leaders appear caught between wanting to ring alarm bells about climate change and a desire to avoid the touchy and seemingly unresolvable issue of migration.

India claims that about 5 million Bangladeshis already are living there illegally while Bangladesh officials say the numbers are wildly exaggerated. The issue is a constant source of tension between the nations. Climate change isn't helping…

India for its part sees climate change bringing multiple threats. Rivers feeding both Bangladesh and Pakistan pass through India but threaten to dry up because of melting glaciers. Meanwhile the country can barely handle demands for resources from its own citizens and argues that it shouldn't have to accept the victims of a problem caused by the industrialised world….

But Cleo Paskal an associate fellow at the Royal Institute of International Affairs in London said the prospect of large-scale migration from Bangladesh represents a real threat to India and it's one the West should take seriously. 'There are a lot of countries that would like to see India weakened' like China and Pakistan Paskal said. She argued that if Bangladesh is destabilised by climate change, Islamic radical elements are more likely to prey on vulnerable communities. That she said could easily lead to more deadly attacks like the one in Mumbai India earlier this year…

The story told in Box 9.2 is interesting because it shows how India has defined a public response to behaviour of citizens of a neighbouring country who face a challenge related to climate change. This policy has been shaped by perceived military and geopolitical threats. More broadly it has been shaped by self-perceived national interests and alarmist assumptions—both of which are fuelled by citizen concerns. All of these and more voices have combined to establish a vigorously enforced 'keep them out' policy.

In this example, no doubt, other voices have cautioned against such alarmist assumptions of a potential security threat. They will have argued that people, no matter how harsh the circumstances, will leave their homelands only as a last resort as hopelessness intensifies. Many Bangladeshis will in fact remain in spite of this and attempt to adapt to their changed circumstances as they do not wish to (or do not think they have the capacity to) develop new social networks, compete for work and housing in already overcrowded neighbouring and urban markets. In fact, some argue (e.g. Popovski and Mundy 2012) that populist claims that huge numbers will migrate by the middle of the century, leading to world chaos and conflict (e.g. Dyer's (2008) *Climate Wars* with 250 million displaced migrants) are ill founded. They even query academic claims based on lengthy research such as that of Couldrey and Herson (eds) (2008) as unconvincing, perhaps too alarmist.

These cautionary voices, however, have been submerged by those that are much stronger as India seeks to make its border with Bangladesh impermeable. No rationalist approach to public policy can cope with this kind of thing, but it illustrates yet again how a public action lens will immediately come to terms with it. Collective lived experience of citizens, in so far as it represents a collective public will, plays a significant part in shaping the policy response.

9.7 Power Politics and Contested Knowledge

Real issues of power underlie a socially-led public action approach. These concern power play around the question of whose knowledge counts, and the political use of contested knowledge to cancel evidence claims. However, we should also recognise that, despite the challenges, difference and diversity are sources of social learning and new knowledge. We do not learn from being the same.

All actors within policy processes are informed by a knowledge base which is used to justify their respective positions. This knowledge base may overlap with that of other actors to the extent that a negotiating alliance is formed, for example, at an international level, the common negotiating position of the European Union countries on climate change is evident (Breitmeier and Otto 2012). Equally, however, the knowledge bases of different actors may be in contention leading to seemingly intractable disagreements.

At root, knowledge is a source of power which may be utilised by actors within a policy process. It is used both to promote policies and interventions, and also to make attempts at blocking them (see Chap. 3). In the UK, for example, the knowledge of anthropogenic global warming is used by the Government as a reason for promoting a renaissance of the country's nuclear power industry.[1] Nuclear power is claimed to be a 'green' energy, as it does not release carbon dioxide or other greenhouse gases. Thus, it is considered a potentially significant contributor to climate change mitigation. Opponents of this renaissance, however, point to the well-documented knowledge of the problems of dealing with the radioactive nuclear waste that is produced in the process, which has been described as the 'achilles heel' of the nuclear industry (Blowers and Sundqvist 2010).

The 1990s challenges about North/South[2] inequality also shook the climate change discourse and its assumptions of factual universal truths and actions. Increasingly questions were asked of science interpretation and its intention both through loud voices from developing countries and the growth of the environmental movement. For instance, in a very early seminal report of the time, Agarwal and Narian (1991: 1–2) fiercely question the purpose of scientific data, arguments which still echo today:

> The idea that developing countries like India and China must share the blame for heating up the earth and destabilising its climate, as espoused in a recent study published in the United States by the World Resources Institute in collaboration with the United Nations, is an excellent example of environmental colonialism.
>
> The report of the World Resources Institute (WRI), a Washington- based private research group, is based less on science and more on politically motivated and mathematical jugglery. Its main intention seems to be to blame developing countries for global warming and perpetuate the current global inequality in the use of the earth's environment and its resources.
>
> A detailed look at the data presented by WRI itself leads to the conclusion that India and China cannot be held responsible even for a single kg of carbon dioxide or methane that is accumulating in the earth's atmosphere. Carbon dioxide and methane are two of the most important gases contributing to global warming. The accumulation in the earth's atmosphere of these gases is mainly the result of the gargantuan consumption of the developed countries, particularly the United States…
>
> The manner in which the global warming debate is being carried out is only sharpening and deepening the North-South divide. Given this new found interest in the so-called 'Our Common Future' and future generations, it is time for the Third World to ask the West, 'whose future generations are we seeking to protect, the Western World's or the Third World's'? The WRI report reinforces this divide. By shifting the onus onto the developing world, it white-washes the role and the responsibility of the West in destroying our 'common future'.

Our intention in repeating this quotation of Agarwal and Narian is not to judge its validity and fairness in relation to the report it criticises, but to consider its tone and overall claims, and the fundamental—ontological—thinking that underlies them. Thus, although the concerns expressed are about scientific data, we cannot disassociate its claims, clouded in emotionality, with other realities—that of the issue of inequality between the North and South which is located in historical relationships of oppression and financial and political inequalities.

---

[1]The other main reason concerns national energy security, so that the UK is not dependent on major oil-producing countries for its energy needs.

[2]I use the terms North and South as these were prevalent at the time. They refer respectively to rich, developed and poor, developing countries.

These other realities have since made an impact in the IPCC 2014 assessment synthesis report which clearly identifies that sustainable climate change mitigation can only be achieved within the context of equity, justice and fairness. It states,

> 'Countries' past and future contributions to the accumulation of GHGs [Greenhouses Gases] in the atmosphere are different, and countries also face varying challenges and circumstances, and have different capacities to address mitigation and adaptation.

The IPCC report further recognises the importance of value judgements and ethical considerations which shape many areas of climate change policy making. To quote,

> These areas range from the question of how much mitigation is needed to prevent dangerous interference with the climate system to choices among specific policies for mitigation or adaptation. Social, economic and ethical analyses may be used to inform value judgements and may take into account values of various sorts, including human wellbeing, cultural values and non-human values (IPCC 2014 Summary for Policymakers: 5).

While we cannot make an explicit association between Agarwal and Narian's fierce attack in 1991 and the recent stance taken in the IPCC assessment synthesis report of 2014, its combination with many other attacks in the same vein illustrate the cumulative weight that eventually finds a mainstream acknowledgement. This weight is not necessarily the power of any specific analysis, which may or may not be valid 'scientifically'. Rather, it is the emotions and discontent that lie behind the attacks. These cannot be dismissed as irrelevant—the United Nations needs to bring the historically oppressed and disadvantaged countries on board for meaningful international action on climate change. They therefore have a source of power. It is a form of public action at work.

9.8 Lived Experience and Public Action Through Demonstration

Despite the examples and argument about the role of lived experience in public action that we have discussed above, collective lived experiences do not always and inherently play an antagonistic role in public policy debates. One paper on social acceptance of renewable energy interventions to mitigate climate change points to several ways in which citizens play an active, generally supportive role. These include as producers, users and investors, as well as opponents, of renewable energy (Fast 2013).

Sometimes, however, individual actions are not simply about working with public interventions (and in that sense legitimising them). They might also demonstrate practical innovations to public authorities with messages so clear that the innovation itself becomes an object in the public action process. They also demonstrate how, on a practical level, experiential and scientific knowledge might combine synergistically.

This is apparent in the following example which has been reported internationally and has generated an intriguing debate on rice growing amongst producers.

It concerns Sumant Kumar, a young farmer from the Nalanda district of India's poorest state, Bihar. Unlike other Indian states such as Punjab, Bihar has not been a beneficiary of India's Green Revolution that started in the 1960s with High-Yield Varieties of rice and associated technological inputs.

Yet in 2012 Sumant grew an astonishing 22.4 tonnes of rice on one hectare of land using farmyard manure and without herbicides, being an increase of almost 30 % on the regional yield. He made world news, challenging attempts at increasing yield by the Chinese scientist and 'father of rice' Yuan Longping, the World Bank and other renowned rice research institutions, as well as European and American seed companies that incorporate genetic modification.

In what is labelled a 'system of rice (or root) intensification' (SRI), the technique combines a 'less is more' innovation with a generational, experiential knowledge of soil and is seen as one of the most significant developments in rice growing in the past 50 years. Yet, it is not that this is new knowledge. This system has been known since the 1980s, but other food technologies (such as Genetic Modification) capable of producing huge tonnage of Basmati and other long-grain rice for export have been dominant. Small producers and their growing techniques have been on the back burner.

Under the guidance from an NGO, Sumant nurtured and transplanted rice seedlings one by one when these were very young instead of the clumps of three to four that most rice growers do. They were also spaced in a larger 25 cm interval grid pattern to keep the soil drier (instead of traditional soggy soil), allowing for weeding around the plants and giving them space to spread roots. Simultaneously, Ravindra Kumar in a neighbouring village dramatically increased yields of wheat, and others have had similar success with potatoes, sugar cane, yams, tomatoes, garlic and aubergines (New record in Bihar; Vidal 2013).

Current developments include a World Bank publication, partnership with American universities and agricultural institutes such as at Cornell University, and knowledge transfer to several developing rice-growing countries such as Ethiopia and Tanzania. Also occurring are practice transfers to other crop varieties such as wheat, pulses, oilseeds and vegetables (Abraham et al. 2014).

Although we have only drawn on one example here, it nevertheless focuses on the very important question of food supply which is threatened in the face of climate change especially in poorer countries. It well illustrates that scientific and lived experience knowledge can work together and may form a strong knowledge alliance in areas of practice. Therefore it is not surprising that India, and indeed the world, has taken note and is looking to an environmental friendly, climate resilient practice of increasing food supply. The example also illustrates that public action need not be an antagonistic affair involving conflict and overt power play.

These types of practical demonstration show that innovation itself can become a part of the public action process and change. Lived experiences that are generated through these processes further suggest how the public understand and respond to climate change at the ground level. This is different (and perhaps stronger) than attempts by science to engage with the public in a top-down, linear, abstract way. We have seen in Chap. 7 that for a significant number of people, the issue of public engagement is not with 'climate change' but with the science of it. This is labelled more broadly within the United Kingdom as 'Public Understanding of Science' (PUS), which is conceptualised clearly within an information deficit model. As one early, influential UK report stated: 'Almost all public policy issues have scientific or technological implications. Everybody, therefore, needs some understanding of science, its accomplishments and its limitations' (Bodmer 1985: 8).

Ultimately, however, the concern of PUS is to legitimise policy through public acceptance and engagement. In other words, it is an attempt to neutralise antagonistic public action that may emanate from citizens, through paradoxically gaining their acquiescence in a linear rationalist policy process that is expert-led. Unfortunately for scientists and government, however, an untrusting public seems intent on resisting being brought on board. They may come to understand more of the science of climate change through PUS initiatives but still fail to internalise it. This is because their lived experiences tell them different things of a more normative nature (Sect. 9.3 above). We see a similar story in areas other than climate change. Examples include medical advances, food technology and so on. There have been serious breaches of trust (e.g. the thalidomide baby scandal, ethics and morality of stem-cell engineering, biotechnology and geoengineering) where scientists are perceived as 'Frankensteins', playing God.

As suggested in the chapter introduction, policy relies on public support for legitimacy and some sense of fair management of tensions between resource priorities, ethical and moral values. Climate change and action around it is too serious to be left to rationalist, linear models which may or may not generate public understanding and participation. A non-linear public action approach, based on individual rationalisation, personal perceptions and lived experience can be more powerful. Taken together, both scientific and lived experiential knowledge can be instrumental in forming a more rounded and robust policy that will touch a public understanding at a wider scale.

References

Abraham, B., Araya, H., Berhe, T., Edwards, S., Gujja, B., Khadka, R., et al. (2014). The system of crop intensification: Reports from the field on improving agricultural production, food security, and resilience to climate change for multiple crops. *Agriculture & Food Security, 3*(4), 1–26. doi: 10.1186/2048-7010-3-4.

Agarwal, A., & Narian, S. (1991). *Global warming in an unequal world: A case of environmental colonialism* (p. 1). New Delhi: Centre for Science and Environment. Accessed February 25, 2015. http://cseindia.org/agenda2011/pdf/global_warming%20_agarwal%20and%20narain.pdf.

Blowers, A., & Sundqvist, G. (2010). Radioactive waste management—Technocratic dominance in an age of participation. *Journal of Integrative Environmental Science, 7*(3), 149–155.

Boykoff, M. T. (2007). Flogging a dead norm? Media coverage of anthropogenic climate change in United States and United Kingdom, 2003–2006. *Area, 39*(4), 470–481.

Boykoff, M. T., & Boykoff, J. (2004). Balance as bias: Global warming and the US prestige press. *Global Environmental Change, 15*(4), 125–136.

Breitmeier, H, & Otto, D. (2012). Understanding political processes in climate change negotiations by means of an interdisciplinary curriculum in higher education. *International Journal of Innovation and Sustainable Development, 6*(1), 20–30.
Couldrey, M., & Herson, M. (eds.). (2008). *Forced Migration Review* (p. 31).
Dankelman, I. (2010). *Gender and climate change: An introduction*. London: Earthscan.
Dessler, A., & Parsons, E. (2010). *The science and politics of global climate change: A guide to the debate* (2nd ed.). Cambridge: Cambridge University Press.
Drèze, J., & Sen, A. (1989). *Hunger and public action*. Oxford: Clarendon Press.
Dyer, G. (2008). *Climate wars*. Toronto: Random House of Canada.
Fast, S. (2013). Social acceptance of renewable energy: Trends, concepts, and geographies. *Geography Compass, 7*(12), 853–866. doi: 10.1111/gec3.12086.
Five ways System of Rice Intensification (SRI) practices and ideas can help 'feed the world'. Accessed February 28, 2015. http://foodtank.com/news/2014/01/five-ways-system-of-rice-intensification-sri-practices-and-ideas-can-help-f.
Friedman, L. (2009). A global 'national security' issue lurks at Bangladesh's border. The New York Times. March 23, 2009. http://www.nytimes.com/cwire/2009/03/23/23climatewire-a-global-national-security-issue-lurks-at-ba-10247.html?pagewanted=all.
Hulme, M. (2009). *Why we disagree about climate change: Understanding controversy, inaction and opportunity*. Cambridge: Cambridge University Press.
Hulme, M. (2013). After Climategate… never the same. In *Exploring climate change in science and society: An anthology of Mike Hulme's writings, interviews and speeches*. London: Earthscan/Routledge.
Hulme, M. (2014). *Can science fix climate change?* Cambridge: Polity Press.
Intergovernmental Panel on Climate Change (IPCC), Abraham, V., Berhe, T., Edwards, S., Gujja, B., & Khadka, B. (2014). *Summary for policymakers* (p. 116). Fifth Assessment Synthesis Report.
Intergovernmental Panel on Climate Change (IPCC) history. https://www.ipcc.ch/organization/organization_history.shtml. Accessed on January 9, 2015.
Jones, S. (2011). *BBC trust review of impartiality and accuracy of the BBC's coverage of science*. London: Imperial College. July 2011. Accessed March 2, 2015. http://downloads.bbc.co.uk/bbctrust/assets/files/pdf/our_work/science_impartiality/science_impartiality.pdf.
Kuruvilla, S., & Dorstewitz, P. (2009). There is no 'point' in decision-making: A model of transactive rationality for public policy and administration. *Policy Science, 43*, 263–287. doi:10.1007/s11077-009-9098-y.
Lasswell, H., & Kaplan, A. (1950). *Power and society, a framework for political inquiry*. Originally published Yale University Press, Available http://www.policysciences.org/classics/power_society.pdf. Accessed 16/01/15.
Lindbolm, C. E. (1959). The science of 'muddling through'. *Public Administration Review, 19*(2), 79–88. Accessed January 16, 2015. http://faculty.washington.edu/mccurdy/SciencePolicy/Lindblom%20Muddling%20Through.pdf.
McManus, K. (2009). The principle of 'common but differentiated responsibility' and the UNFCCC. *Climatico* Special Features. November 2015. Accessed January 19, 2015. http://www.climaticoanalysis.org/wp-content/uploads/2009/12/kmcmanus_common-responsibilities.pdf.
New record in Bihar: Thanks to SRI. Accessed February 28, 2015. http://www.agriculturesnetwork.org/resources/extra/bihar-sri.
Norgaard, K. M. (2011). *Living in denial: Climate change, emotions and everyday life*. Cambridge: The MIT Press.
Painter, J. (2011). *Poles apart: The international reporting of climate change scepticism*. Oxford: Reuters Institute for the Study of Journalism.
Poortinga, W., Spence, A., Whitmarsh, L., Capstick, S., & Pidgeon, N. F. (2011). Uncertain climate: An investigation into public scepticism about anthropogenic climate change. *Global Environmental Change, 21*, 1015–1024.
Popovski, V., & Mundy, K. (2012). *Sustain Science* 7(5), 16. doi: 10.1007/s11625-011-0138-0.
Srinivasan, A., Ling, F. H., & Hideyuki Mori, H. (2012). *Climate smart development in Asia*. London: Routledge.

System of rice intensification: SRI use in other crops. Accessed February, 28 2015. http://sri.ciifad.cornell.edu/aboutsri/othercrops/.

Talking Climate: The Gateway to Research on Climate Change Communication. Accessed January 16, 2015. http://talkingclimate.org/wp-content/uploads/2011/12/Climate-change-scepticism-and-the-media.pdf.

Towards a climate change agreement: http://www.un.org/climatechange/towards-a-climate-agreement. Accessed February 24, 2015.

United Nations General Assembly. 70th plenary meeting A/RES/43/53.1988. December 6. *Protection of global climate for present and future generations of mankind.* http://www.un.org/documents/ga/res/43/a43r053.htm. Accessed January 19, 2015.

United Nations Millennium Assessment Report of the Secretary-General. (2000). *We the peoples: The role of the United Nations in the 21st century.* United Nations Department of Public Information. Accessed January 9, 2015. http://www.unep.org/ozonaction/AboutTheBranch/MDGs/tabid/6185/Default.aspx.

United Nations Framework Convention on Climate Change. Accessed January 09, 2014. http://unfccc.int/timeline/.

United Nations and Climate Change: Towards a climate change. Accessed January 09, 2015. http://www.un.org/climatechange/towards-a-climate-agreement/.

Vidal, J. (2013). India's rice revolution. *The Observer.* February 16, 2013.

Walter, B. (1985). *The public understanding of science*. London: Royal Society.

Ward, B. (2008). A higher standard than 'balance' in journalism on climate change science. *Climatic Change, 86*, 13–17.

Whitmarsh, L. (2011). Scepticism and uncertainty about climate change: dimensions, determinants and change over time. *Global Environmental Change, 21*, 690–700.

Woodhouse, P. (1992). Environmental degradation and sustainability. In T. Allen & A. Thomas (Eds.), *Poverty and development in the 1990s*. Oxford: Oxford University Press.

Woodland Trust. https://www.woodlandtrust.org.uk/. Accessed March 6, 2015.

Wuyts, M., Mackintosh, M., & Hewitt, T. (1992). *Development policy and public action*. Oxford: Oxford University Press.

Chapter 10
Institutionalising Lived Experience in the Public Policy Process

Abstract Reasons for institutionalising lived experience in public policy processes on climate change vary, depending on whom one asks. Government, the private sector, scientists, engineers and civil society each have different, albeit sometimes complementary and sometimes overlapping, reasons. We, the authors of this book, also have reasons, namely that institutionalisation of lived experience within the policy process:

- Exposes the politics and power relations at play in public policy making to those whose lived experiences are being institutionalised.
- Generates a space for joint learning and broader, deeper knowledge alongside the inputs from science
- Exposes the dominant frames within which climate change debates are constructed
- Provides a peg for voicing other pressing social issues that could lead to a polycentric framing

  We also argue that through our institutionalised engagement with one another we come to appreciate our human and human–nature interdependence and the potential to develop a transboundary, social imagination among citizens and scientists alike. Institutionalisation, however, raises many issues. These range from basic human attributes of just how much knowledge we may hold and tell to establishing the basic conditions for productive engagement and managing enduring power relations between actors and their sources of power. The chapter ends by examining three potential ways in which we might institutionalise lived experience in public policy making on climate change:

- Government monitoring of the social media sites that exist on the topic
- Government-organised focus groups and interviews
- Independent environmental action groups as have been established in Southern Africa by an NGO acting as a boundary organisation to bring civil society, scientists, the private sector and governmental actors together.

D. Abbott and G. Wilson, *The Lived Experience of Climate Change*,
DOI 10.1007/978-3-319-17945-2_10

10.1 Introduction

Institutions may be broadly defined as the behaviours and practices that relate to the overall 'rules of the game'. These rules may be enforced formally through the law or informally by expected norms that derive from collective values. Institutions may be organisations, but not all organisations are necessarily institutions. Examples of institutions include public organisations in Europe which provide welfare services such as health for all. They may also be what are termed 'social institutions', which are established aspects of wider society that embody common values and ways of behaving (Thomas 2009: 181). Examples of social institutions include that of marriage and, to take an example closer to the topic of climate change, the informal rules that regulate behaviour of tropical forest communities so that they do not engage in a 'free-for-all' process of forest exploitation (Ostrom 1990). In summary, institutions regulate formally or informally practice and behaviour.

We have argued in Chap. 9 that making climate change policy is, to a large extent, an unruly, political process of public action. It is unruly because machinations of power dominate at every scale. This is despite attempts to regulate through international institutions such as the United Nations and its Framework Convention on Climate Change (UNFCC), and despite in many countries there being opportunities in planning legislation to debate and consult over proposed local interventions. This chapter, therefore, explores the potential of making the process less unruly through institutionalising knowledge that is derived from lived experience, that is, accommodating such knowledge within the 'rules of the game'.

We have explored in earlier chapters throughout this book, but especially in Chap. 8, the productive potential of engaging the sciences and lived experiences through disrupting the boundaries that limit both. Thus we have argued that:

- Socially organised denial which leads to a disinclination of the public to know about and act over climate change is itself a symptom of the limiting boundaries we place around locally produced lived experience through social norms of attention.
- Potentially restrictive boundaries also apply to all other forms of knowledge that inform the climate change debate and policy, not least those of the scientific disciplines that are bounded by their distinctive conceptual frameworks.
- Although restricted in itself, the knowledge contained within one boundary potentially expands and enriches the knowledge contained within another boundary, which at its most productive will generate a transboundary, social imagination.

Fundamentally, therefore, we have argued that difference between the forms of knowledge that are contained within these boundaries is the basic resource for generating an expanded knowledge and a social imagination. In previous chapters, however, we have often noted the other side of the coin. There, difference has been characterised as a power relation between actors in the climate policy making process, where the policy that is eventually made is largely an outcome of those power relations.

Our context for this chapter is the public sphere of policy making. Specifically, therefore, we are exploring the potential institutionalisation of lived experience within the overarching frame of national government and its delegated authorities (for example local municipalities). A key thread running through the chapter is the above tension between difference as resource for an expanded knowledge and social imagination, and difference that embodies power relations between actors, and the ways in which institutions might manage the tension.

We start in Sect. 10.2 by addressing the question: 'Why should we wish to institutionalise lived experience in policy debates about climate change?' Then, in Sect. 10.3 we explore more deeply some of the issues associated with institutionalising it. Finally, to the extent that our answer in Sect. 10.2 is positive despite, or because, of the qualifications highlighted in Sect. 10.3, we ask in Sect. 10.4: 'How might we institutionalise lived experience?'

10.2 Why Should We Wish to Institutionalise Lived Experience in Public Policy Debates About Climate Change?

We cannot divorce this question from another one that asks, 'Who might wish to institutionalise lived experience in climate change policy debates?' Different actors will have different interests in, and hence reasons for, its promotion. Here we outline six groups of actors, the final group being very narrow compared to the previous five:

- Government and its national and local institutions
- The private sector
- Scientists from across the social and natural divides
- Engineers
- Civil society, that is, citizens and groups of citizens who might wish to engage in the public sphere over climate change. This might be for a variety of underlying reasons, for example as activists, as concerned individuals or as communities directly and possibly adversely affected by climate change interventions
- Ourselves, the authors of this book.

10.2.1 Government and Its Local Institutions

Governments, especially those that are democratically elected but to an extent all governments, need popular legitimacy for their policies on climate change. Whatever the claimed *eventual* economic benefits that are claimed for climate change mitigation and adaptation, citizens are more interested in what is happening to them in their *present* circumstances (Chap. 6). Government has therefore to take account of current impacts, not only of climate change itself but also of interventions that are taken in the present to address the challenge. This situation might be labelled short-termism and lamented by political commentators but it is a reality that, for the sake of popular legitimacy, cannot be ignored.

Thus, national governments at an international United Nations climate change event have to keep an eye on the public mood back home when discussing coordinated international intervention that will require financing or might add directly to increased energy costs for consumers—for example a price on carbon dioxide emissions. Governments also have to take note of any outcry that might arise when, for example, they apply a fossil fuel energy tax to suppliers as part of a policy to stimulate innovation in, and diffusion of, carbon neutral alternatives. This has happened in the United Kingdom, for example, because the tax is then passed from suppliers to consumers—the public—who in turn blame the Government for the rising cost of domestic energy. As we write, a general election in the UK is about 6 months away, and all of the main political parties are committing to reducing fuel bills where the fossil fuel energy tax is an easy target to attack.

The prime reason why Government and its local institutions might wish to institutionalise lived experience, therefore, is for purposes of what the philosopher Michel Foucault termed 'governmentality', a term that we introduced in Chap. 6. For Foucault (1979), governmentality is the art of modern government and requires knowing one's population—its needs and characteristics, its wants and desires, what is acceptable and what is unacceptable—in order to make it governable. Specifically, for our purposes, the population becomes governable in relation to climate change when it provides legitimacy to government policy and actions, at least to the extent that it does not object strongly to them (Chap. 6).

Therefore, Government has to learn about its population (Chap. 5) and for this it requires a data set of everyday impacts of, perceptions of and reactions to its policies. Such a data set may be gained through knowing about the population's lived experiences. Statistical gathering, classifying and mapping methods may be employed, where quantifiable proxies are used to represent lived experience, such as opinion or acceptance surveys. Interviews and focus groups may add a qualitative dimension that potentially captures more directly some of the major issues that arise out of lived experience. We address interviews and focus groups in Sect. 10.4 below.

The legitimacy of government intervention and policy on climate change has two general dimensions: (i) the popular legitimacy of the policies and interventions that are eventually decided, which is the more obvious dimension that we have assumed above; (ii) the legitimacy of the governmental policy process in reaching its decisions. Thus, in relation to the latter, Government might also wish to institutionalise lived experience as a form of knowledge in order to expand the evidence base of, and give credence to, the so-called policy cycle of rational, technical, evidence-based decision-making.

Within the process dimension, lived experiential knowledge and other (professional) evidence are then arguably required at all stages of the policy cycle, for:

- Defining the rationale and objectives of the policy
- Appraising different options for achieving the objectives
- Monitoring and evaluation during implementation to provide practice-based evidence regarding the extent to which the policy is actually working. Evaluation may result in minor adjustments that are needed, pointers for alternative options beyond minor adjustment or even a different policy.

The policy cycle for the United Kingdom Government Treasury (Economics and Finance Ministry)—ROAMEF—is shown in Fig. 10.1 and is fairly typical of many around the world. It is reproduced from a report by Hallsworth et al. (2011: 25) for the UK Institute for Government—an independent think tank that promotes more effective government.

As noted above, evidence is required for each of these stages. Hallsworth et al. (Ibid.), however, provide a critique of this rationalist approach to policy, concluding:

> In this 'ROAMEF' cycle, each stage follows on rationally from the previous one, so that a rationale is developed, then objectives are set, then options are appraised. The ROAMEF cycle presents policy making as a controllable sequence where 'the government' produces a 'policy' that addresses a clear goal. The policy represents a set of planned actions that are then implemented, with monitoring to assess the extent to which the goal was fulfilled. The framework is technocratic, with politics, values and events seen as external 'noise' that needs to be minimised.

The 'ROAMEF' cycle represents a rationalist approach as applied to policy, which may be termed 'policy as prescription', that is, it is formulated by and handed down from government. You, the reader, will note from reading Chap. 9 that, unsurprisingly, we take a similar view to that of Hallsworth et al. of this approach which attempts to place it outside of politics. For us, it simply does not correspond with real-world policy processes. In recognition of this, Hallsworth et al's report is titled 'Policy making in the real world' in order to contrast its argument with the neat diagram of Fig. 10.1. The policy cycle is, therefore, a myth, or at best aspiration, of evidence-based rationality and technicality that governments all

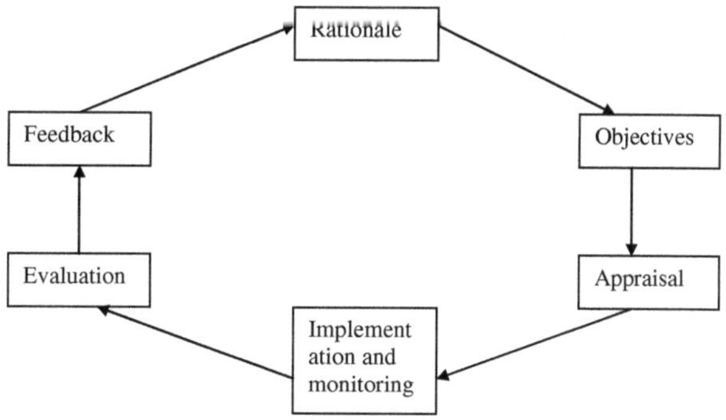

Fig. 10.1 The policy cycle (ROAMEF) of the UK government Treasury

over the world wish to promote to show that they are in charge and capable of wise decisions. As a forthcoming paper in the Journal of International Development (JID) states, the rationalist policy cycle facilitates 'the circulation of a "public transcript" which promotes the idea of measured and thoughtful practice'.[1] It is to reinforce and provide authority to this public transcript that government may wish to incorporate evidence garnered from analysis of lived experiences.

Beyond the above considerations, there are undoubtedly many specific aspects of the governmentality argument that provide particular reasons why Government might wish to institutionalise lived experience into policy making. Here are two examples to end Sect. 10.2.1.

First, climate change is arguably affecting people adversely in many parts of the world now in terms of rising sea levels and weather variability. Evidence from physical science, however, deals primarily with what happens over time on a generalised scale. It concerns to a lesser degree particular, current events where, with respect to weather, scientists insist that they are unable to relate any single event to climate (Chap. 3). Thus, the headline concern and public bias of climate physical scientists is to produce scenarios of a future where the earth continues to warm by x, y or z degrees, with projected impacts of p, q and r. As a correction to this bias, Governments have good reason to institutionalise narrations of lived experiences. They provide a reality check on what is happening now, including the impacts of current attempts to mitigate and adapt. That is what articulation of lived experience is primarily—making sense of the present and projecting it into the future in terms of our experience to date.

Second, especially in poorer countries, institutionalising lived experience is a way of finding out about local 'below the radar' innovations that vulnerable people and communities across the world are making to adapt now to climate change that is already affecting them. Meanwhile, government institutions move slowly where innovation is not their strong point, while the private sector is much faster at innovating in response to market signals but has a more major underlying motive in the form of maximising profit. In this context, community-based adaptations (Chap. 5) are a useful demonstration of what is being done in the present, usually on low budgets. In effect, they are about innovating to achieve more from less.

10.2.2 The Private Sector

As sellers of goods and services, private sector firms, big and small, are seen everywhere by governments, supranational bodies such as the European Union and

[1]This paper was reviewed by one of the book authors and has since been accepted for publication, but is unlikely to be published before this book appears. It is titled 'Knowing and Doing Between the Academic and Policy Worlds in Development Assistance' (JID editor, private communication).

mainstream economists to play a key role in climate change policy implementation. This is for two main reasons.

First, by definition the private sector consumes a great deal of energy and therefore collectively creates a large carbon footprint, which needs to be brought down. Second, it constitutes the main means of delivering carbon-friendly technologies, goods and services for the rest of us. In short, private sector firms need to be incentivised to innovate. The favoured incentive, at least of the affluent countries, is to put an international price on carbon emissions to correct for the 'market failure' of climate change (Chap. 6, Sect. 6.3). For example,

> If public actions and policies give the right signals and rewards for cutting greenhouse gases, then markets and entrepreneurship will drive the response. The bulk of the action will be in the private sector – this is not about a return to government control and rigid planning: on the contrary it is about enabling markets and private-sector initiative to work well.
>
> (Stern 2010: 99).

Thus, the private sector, like Government, will have an interest in knowing where the public is really at because:

- It needs to know the public as consumers and what they are prepared to do (and maybe sacrifice) in terms of purchasing carbon-friendly products and services.
- Companies will gain greater clarity of public issues and the political difficulties that lie ahead. This might seem counter intuitive—surely institutionalising knowledge of lived experience in government will muddy the already murky waters of policy making still further, it will be argued. However, such knowledge might actually help clarify government policy, just as it helps clarify marketing of products. In the language of business, the policy cycle will potentially be a more efficient process. It will also generate some useful market research findings.

10.2.3 Scientists Across the Social and Natural Divides

Why should scientists of all subjects and disciplines that have an input to climate change debates wish to institutionalise lived experience as a legitimate knowledge input to the policy making process? Paradoxically, it potentially creates a space for scientists themselves to communicate their knowledge to members of the public. Many now understand the limitations of older approaches, such as the knowledge deficit model which posits that the public does not engage with climate change because of lack of information and understanding that results from poor communication of the science. Consequently, they are looking to more dialogic approaches in order to put their message across. For example, the edited volume of Jaeger et al. (2012) that we discussed in Chaps. 6 and 7 was written by 17 natural and social scientists and from their scientific perspectives as it sought to move the debate on climate change communication towards finding 'win–win' solutions.

Engaging in dialogue with the public, however, means acceptance of the validity of its experiential knowledge. Engaging in that knowledge may also help scientists understand the increasing public mistrust of science (see subsection on civil society below) to provide answers, and to win back that trust as earned. A further gain for scientists is that engaging with lived experiences will bring them greater understanding of the policy process on climate change. Above all, it will help them to understand that the policy process is not a matter of simply enacting the logical conclusion of the science.

Beyond these general observations, there are of course many disciplines of science that might, or might not, have a particular subject interest in institutionalising lived experience as a legitimate form of climate change knowledge. Thus:

Physical scientists who work on climate change are unlikely to have a direct subject interest in lived experiences. They deal with energy balances and transfers, circulation of air and ocean currents, and absorption properties of atmospheric gases.

Life scientists are concerned with the impact of climate change on biodiversity and ecosystems of plants and animals, but not directly on humans (although these changes will in turn affect humans). They too are unlikely to have a direct subject interest in lived experiences.

Psychologists who work on climate change are interested in human behaviour in relation to the challenge. They potentially, therefore, have a subject interest in lived experiences, although many are primarily concerned with behaviour with respect to the cognitive functions of the brain rather than the social influences on behaviour.

Social scientists of all connected disciplines have a potential subject interest in lived experiences as they explore the known and probable economic and social impacts of climate change, its differentiation between social and geographical groups, and the societal responses to it. Sociologists and anthropologists will have a special direct interest—some may even be researchers of lived experience if not of climate change, then of other contexts, for example of poverty and of health care.

10.2.4 Engineers

Engineers and physical scientists are often put together for what they share, which is a foundation in scientific principles. Their respective worldviews are, however, different in important aspects and the reason why engineers might like to see lived experience institutionalised in climate change discussions is likewise different. Thus, physical scientists by definition are concerned with the physical, not the human, world. Their research may lead to practical application eventually, but their daily work is usually outside of that context. Engineers, on the other hand, work at the sharp end of practice, they are the practical problem solvers, the ones who will ultimately design and implement the practical interventions on climate change—the clean energy technologies, processes and products.

This difference means that engineers tend to appreciate the professional challenge and gain professional satisfaction from working in the 'real world' to solve problems, away from the laboratory that is beloved of physical scientists. The 'real world', however, is full of people as well as 'things', people whom they have to engage repeatedly in their daily work. Engaging with lay people and their lived experiences in turn forces them to ask fundamental questions of themselves concerning why they do what they do.

Reflection on these considerations takes them back to the roots of their profession. For example, Johnson and Wilson (2007, 2009a), Wilson and Johnson (2007) interviewed British environmental health engineers who were in partnership with Ugandan counterparts. The work of these British engineers in Uganda forced them to 'throw away the book' and return to first principles. One interviewee said that the experience took him back to the great public health works in water and sanitation in the nineteenth century that were associated with a large drop in the incidence of disease and child mortality in the United Kingdom (Wilson 2007, 2008). Their engagements with lay lived experiences during a consultation exercise with Ugandan citizens and other professionals also induced a degree of humility, as they realised that this is not simply an 'engineering problem and solution, and this is how we're going to do it… being able to get local people to support us—the community, the disabled… This is a better way and it questions the whole foundations on which you stand' (Wilson and Johnson 2007).

10.2.5 Civil Society

By Civil Society we mean the public that engages with the world beyond its own private interests. In practice it refers largely to active citizens and citizen groups who, in the process of engaging with the world, inevitably bring their own lived experiences to bear.

Why should active citizens and citizen groups wish to institutionalise their lived experiences as legitimate knowledge inputs to the public policy making process on climate change? Here are five interrelated reasons:

1. The increasing mistrust of politicians, scientists and other expertise, at least in affluent countries, to arrive at rational, objective and disinterested decisions on behalf of the population at large (Hulme 2010a; also Chaps. 3 and 4). Institutionalising lived experience provides the space, therefore, for citizens to confront through a different form of knowledge, politicians, scientists and other experts who currently provide inputs to the policy process. It challenges the once-ascribed authority of these people and demands that any authority is henceforth earned.
2. Active citizens might feel the need to add a humanly felt, perhaps emotional, dimension to give meaning to societal impacts of climate change as explored by the social sciences.

3. They might see opportunity to connect lived experience of climate change to their experience of other pressing social issues such as poverty, inequality and vulnerability. In other words, institutionalising the lived experience of climate change potentially enables active citizens to link to, and press the case for, addressing other real contemporary problems in tandem, as advocated by, among others, Hulme (2009, 2010b), Jaeger et al. (2012). See also Chap. 7.
4. As we saw in Chap. 5 and in Sect. 10.2.1 above, many communities in poorer countries that are vulnerable to climate change already have to innovate and make their own adaptations with whatever resources they have. Institutionalising lived experience of climate change demonstrates their agency, that with appropriate resources they are able to act to help themselves. It may also help lever greater financing for poor communities to innovate and create their own adaptations, which we saw in Chap. 5 was a key topic of the 2014 Community-based Adaptation annual conference in Nepal.
5. The fifth, more general, reason that active citizens and citizen groups might wish for government institutionalisation of lived experience is that it provides a space for political empowerment within the policy making process. The knowledge they bring of lived experience is legitimised and their views are sought. Critics may argue (see Sect. 10.3 below) that such empowerment is illusory, but it is not as clear-cut as that. Besides, whether illusory or not, that active citizens and citizen groups might simply *view* institutionalisation as empowering could be one of the benefits they at least perceive.

10.2.6 Us, the Authors of This Book

We should state at the outset that we do not reject any of the possible reasons for institutionalising lived experience that we have attributed to different groups above. This may appear strange at times. For example, that experts from Government and elsewhere might wish to extract data from analysis of lived experiences of populations for purposes of governmentality (Sect. 10.2.1) is not well received in much academic literature. Implied is that extracting data in this way is exploitative. In relation to citations that have already appeared in this book, see Chap. 5 and references that concern data extraction within participatory practices, notably: Chambers (1997: 111, 214), Cooke and Kothari (2001), Craig and Porter (2003) and Rakodi (2000).

For us, however, whether or not data extraction is in reality a feature of exploitive relationships depends on various factors. These include: the feelings of the target population about being used for data; feelings about the transparency of those who extracted it concerning their motives; how the data is used; and whether those from whom data has been extracted have been able to gain knowledge in return. We are not, therefore, seeking perfection in these matters. Most important to us are the spaces that are opened by institutionalising lived experience whatever the motives of government or other actors.

Below are the distinctive reasons why *we* favour institutionalising lived experience in policy making. They sometimes overlap with, while going beyond, those for other actors that have been suggested above.

Thus, we suggest that institutionalising lived experience in policy making opens a space that potentially:

- Exposes the politics and power relations at play to those whose lived experiences are being institutionalised. Citizen participants in the process start to realise how power and knowledge work in these situations. Realisation is the first step towards challenge and reclaiming the power of citizens to set the public agenda with respect to climate change. Meanwhile, the very institutionalisation process accords their lived experiential knowledge some authority and hence power. Citizens understand that their consent is needed and that no power is absolute, equally no powerlessness is absolute. They come to understand, by direct experience, policy making as a political process.
- Includes the diversity, not just of scientific forms of knowledge on climate change, but of non-scientific, 'common sense' forms. From inclusion of diversity we may then move towards exploring it as a resource for joint learning and broader, deeper knowledge. The next step then would be to institutionalise engagement between different epistemic groups—between diverse lived experiences, between diverse scientific disciplines, and between lived experiences and the sciences. From there we may also move from learning *from* one another (which leads to mutual data and information extraction) to understanding one another and learning *with* one another (Wilson 2006, 2007, 2008).
- Exposes the dominant frames within which climate change debates are constructed. Examples of current frames include:
  - climate change as an extended battle or war that requires sacrifice;
  - climate change as representing a catastrophic market failure where efforts must be focused on obtaining an international agreement to put a price on carbon emissions;
  - the public as having little will to know and act on climate change (see next bullet).
- These and other frames were presented in Chap. 7. Foucault (1980) called such overarching frames 'regimes of truth' or dominant discourses within which debates are conducted. Debates-within-frames could, for example, be over the acceptable level of sacrifice by current generations that is needed to win the war; the incentives that are needed in development aid to poor countries to bring them on board; and the lessons we might learn from the media or marketing departments about how to engage the public.
- Exposure, however, potentially breeds challenge to these overarching frames/regimes of truth/discourses. Foucault (Ibid.) pointed out that they do not just emerge, they come from somewhere. They are in fact constructed out of a complex negotiation between actors, and the relations and sources of power between them, to arrive at a 'correct' narrative that represents the truth.

- Key to establishing the current national climate change frames/regimes of truth/discourses have been government, the energy sector of the economy, physical scientists and mainstream economists. Yet we should not consider them as inviolable, they have evolved from something and they will doubtless evolve into something else. For example, in the European Union from about 2007 a dominant frame/regime of truth/discourse has been that tackling climate change can bring positive impacts to the regional economy in terms of jobs, prosperity and energy security (that is, its energy sources not being beholden to foreign oil exporters from Russia and the Middle East). This truth/discourse/frame then creates an imperative: the European Union must modernise its economy towards one that is not so energy intensive and concentrate instead on developing renewable energy and green products, while investing in the necessary human skills. It is a truth/frame/discourse that can be seen in part as trying to win over a public to climate action through accommodating its needs and expectations, at least in terms of public rhetoric. It seems obvious today, but was not so in the early 1990s when it began evolving. We suggest that one reason for this was the early failure to make explicit the links between climate and the economy which we take more-or-less for granted today. Although, in 1990, the European Union and country governments would have known about their populations' concern for jobs, they would not have linked it to climate change.
- In other words the truth/frame/discourse shift has been made in relation, at least in part, to the EU accommodating the population's core needs. Moreover, simply to make this accommodation, which is tantamount to recreating subjective reality, indicates a kind of diffuse, if often latent, power among the population.
- Challenges especially (for this book) the basis of the 'public will to know and act' frame which we explored in Chap. 7. This frame ultimately places failures of individual citizens as the central problem where either we do not know because of inadequate communication of what the science has made clear about climate change, or we do not want to know because of socially organised denial. Hence we resist taking significant individual actions beyond limited consumer choices related, for example, to lowering individual carbon footprints, and are often antagonistic to government actions that concern climate change mitigation where this might affect us directly and negatively. Institutionalising lived experience, however, does not place the public and our limited knowledge as the central problem. Rather, the failure is one of not *recognising* lived experience as a legitimate form of knowledge, not *understanding* the epistemological limits of lived experiences *and* the sciences, and lack of meaningful *engagement* across the different forms.
- Provides a peg for voicing other pressing social issues that could lead to a polycentric framing of the type advocated by Hulme (2009, 2010b) and others. We saw in Chap. 1 how it is in practice impossible to separate lived experience of climate change from these other issues. To do so might work as a heuristic device, for analytical purposes, but to most people everything is interconnected through their lived experiences. Those whose lived experiences are being institutionalised will not put a neat boundary to isolate the bit that says 'of climate

change'. They will raise these other issues that may be of greater importance to them personally. In doing so they will challenge the climate change frame of '*the* war' of the twenty-first century. This reframing will be further reinforced, moreover, if people with very different lived experiences engage over social divides, where citizens come to understand the standpoints (Chap. 9, Box 9.1) of different social groups locally, nationally and internationally.

The above combine to form two overarching reasons that we advocate for institutionalising lived experience into policy making and intervention:

1. Through our institutionalised engagement we come to appreciate our human and human–nature interdependence (Smith et al. 2007) with respect to climate change and other social issues, where nobody by definition is absolutely powerless. From appreciation we may then construct the substantive meaning of our interdependence while avoiding deterministic exhortations such as: 'We must stand together as a human race to fight the war of climate change'. Rather, through engagement of lived experiences across the world, we might construct new frames (for example the polycentric frame as above) for climate change and allow for creative meanings and even disagreement to evolve. We might also come to appreciate, not just the complementarity, but also the interdependence of different forms of knowledge in order to create new knowledge. We may construct these creative responses to interdependence through 'learning with' and empowering each other rather than the disempowering effects that the impending disaster narrative and its imperatives force on us (Ibid.).
2. We potentially challenge the status quo on climate change knowledge through developing a transboundary, social imagination among citizens and scientists alike. We have argued wherever pertinent in this book that the status quo, providing insights especially from the physical sciences, economics and sociology, has brought us a long way. It is not sufficient, however, to capture a full engagement with climate change, nor does it in the final analysis capture the human dimensions in terms of responses and impacts on lives.

For us, therefore, institutionalising lived experience in policy making provides a space for broad engagement. We are broadly in line with Cornwall's (2002) description of spaces as opportunities, moments and channels where people can act to potentially affect policies, discourses, decisions and relationships that affect their lives and interests. Note the word 'potentially' in this description. While we are reasonably sure that they will exist, we are very unsure of what will take place in such a space. It is unpredictable and thus here we have presented only potential and positive possibility. One way of encapsulating the uncertainty that surrounds space is to think of the issues that we present in Sect. 10.3 below as one end of a spectrum, and the enactment of positive potential and possibility that we present here as the other end. Section 10.4, meanwhile, suggests forms of engagement that may provide the conditions for moving towards the positive end of the spectrum.

The starting question for this Sect. 10.2 was: 'Why should we wish to institutionalise lived experience in policy debates about climate change?' Having been through the diverse reasons that we have postulated for different groups, our summary answer is: it depends on who you ask. That answer might seem unhelpful, but a start has to be made somewhere. To move forward, everyone has to agree on one thing, which is that there is sufficient justification across a range of stakeholders for institutionalising lived experience as a legitimate form of knowledge in public policy debates about climate change. This provides a conceptual container within which the diverse interests of actors may be accommodated (Isaacs 1993; see also the same brief reference in Chap. 7) while recognising that, substantively, the different reasons for institutionalising lived experiential knowledge might often be complementary, but also sometimes in conflict and not made clear.

Conflicting reasons among stakeholders bring us to Sect. 10.3 and the issues and the problems that may arise through institutionalising lived experience in policy debates about climate change.

10.3 The Issues (and Problems) Associated with Institutionalising Lived Experience in Public Policy Debates About Climate Change

Some issues represent the opposite side of the coin to the potential that we suggested in Sect. 10.2. Others are different. Here we separate the issues (and problems) into three groups: issues of 'getting real'; issues of creating the conditions for productive engagement with and between lived experiences; and finally the issue of social power even when lived experiences are institutionalised within climate change public policy debates.

10.3.1 'Getting Real'

In its ideal form, the space that opens when lived experience is institutionalised in policy making enables diverse lived experiences to engage with each other across social divides. Given that scientists also have lived experiences it also helps them to engage with each other across their disciplinary divides, and for the diverse lived experiences of citizens to engage with the diverse scientific disciplines. In practice, however, this plurality of engagement is almost certainly unrealistic, and probably impossible, to achieve. The logistics alone of both launching and sustaining such a space appear formidable. Even if it did occur, analysing and drawing together useful findings would represent a further major challenge.

Moreover, we can only *know* so much. The Harvard professor, Atul Gawande, commented in a 2014 interview with the London-based *Financial Times*: 'The knowledge that exists to help people exceeds what you're going to be able to hold in your head' (Gawande 2014). This statement, moreover, constitutes the first thing that he would teach future doctors on their first day of class, but we may adapt it just as appropriately to climate change knowledge. Thus, the knowledge that we need to respond appropriately to the challenge of climate change exceeds what we are going to be able to hold in our heads.

We also can communicate verbally only so much, or as Michael Polanyi (1958) put it: 'We know more than we can say'. This is because much of our knowledge is tacit, that is, it is felt rather than written down or codified, and is not easily passed to others. For Michael Polanyi, much of lived experience would be classified as tacit knowledge.

There are, therefore, inherent human limits to what we can both know and say, and we have to work within those limits. These observations in part explain why policy makers tend to prefer their evidence in the form of numbers (Adamson 2013). Numbers are a way of reducing complex factors embodied in evidence so that they are easily comprehensible and inform generalisations out of which policy and intervention may be specified (Abbott and Wilson 2014). They inherently categorise and therefore simplify complex phenomena, such as human responses to climate change. Lived experiences, however, require qualitative methodologies (Chap. 4, Sect. 4.5.1) as so much would be lost if we tried to turn them exclusively into numbers.

There is a further problem in that an 'all singing, all dancing' space that generates and captures knowledge qualitatively to inform public policy will be of necessity associated with a participatory forum for direct democracy. We came across in Chap. 5 critiques of participatory processes as constituting a 'new tyranny' (Cooke and Kothari 2001). In that chapter we also quoted Leroi's (2014: 316) critical description of direct Athenian democracy in the fourth century BC as 'institutionalised mob rule' where 'demagogues roused the rabble'. The point is that participatory direct democracy cannot escape from issues that relate to power relations and representation/non-representation of interests. The conclusion which we draw, therefore, is that it might be better to create and invite a limited number of cross-sectional representatives of scientists and citizens who represent a range of interests and social groups to the communication table, rather than promote an all-in space.

10.3.2 Creating the Conditions for Productive Engagement with Lived Experiences, and Between Lived Experiences and Other Forms of Knowledge

The Ancient Greek philosopher, Aristotle, who we pursued relentlessly in Chap. 3 as we analysed the origins of science, said in one of his works: 'For the things we have to learn before we can do them, we learn by doing them' (cited in Bynum and Porter 2005).

Many centuries after Aristotle, 'learning by doing' became a focus of pioneering work of Kolb (1984) and others on experiential learning. Through 'learning by doing' we also gain our tacit dimensions of knowledge as espoused by Michael Polanyi (see above). For this book, it means that true engagement is more than simply talking with each other. Engagement also means shared experience and shared practical application, of doing together (Weick Karl 1995: 188–189). For example, in the research that Johnson and Wilson conducted on partnerships between municipal environmental health engineers in the United Kingdom and Uganda, one British engineer commented that that there was nothing better for deepening bonds with your counterpart than standing in a storm drain together and getting your feet wet and muddy (Johnson and Wilson 2009b). We suggest further that joint practice can be an important leveller of hierarchical differences between theoretical and experiential knowledge (Wilson 2007), which is important when we discuss the power issue below.

This comment about joint practice also introduces a related condition for learning. Thus, the shared experience in Johnson and Wilson's research deepened bonds and in a real sense deepened trust between the two engineers. Generalising from this observation, we suggest that trust between the actors at the level of personal micro-engagement is essential for productive interchange of different forms of knowledge. Under such conditions actors are more prepared to expose themselves before each other, push the boundaries of what they know, explore radical ideas together, and embrace disagreement where necessary.

Shared experience and trust between actors are, therefore, important conditions for genuine dialogue. There are, however, significant issues in establishing both. Shared experience requires repeated engagement and reinforcement for it to be meaningfully established, and there are serious practical difficulties in establishing this. People, be they scientists or citizens, have many other things happening in their lives and cannot devote all of their time engaging repeatedly and reinforcing their shared experience. On top of that, developing trust is no easy task where there are likely to be deep-seated ontological differences between the actors. While an international network of scientists of a particular discipline might see the world in broadly the same way, and similarly an international partnership of problem-solving local government engineers will do so, this is unlikely to be the case when, for example, scientists and locally based civil society groups attempt to relate to each other (Ibid.).

Many commentators indicate other conditions for productive, dialogic engagement that relate to personal attributes. For example, Regan (2007: 2956) echoes other literature when she cites a United States action research project—the Public Conversations Project—where 'good dialogue' offers participants the opportunity to:

- Listen and be listened to so that all speakers can be heard
- Speak and be spoken to in a respectful manner
- Develop or deepen mutual understanding
- Learn about the perspectives of others and reflect on one's own worldview.

There is again an issue, however, as to whether being able to listen, be respectful to others, try for mutual understanding, learn about different perspectives and reflect are attributes that are ascribed from birth or alternatively are the result of lifelong social conditioning (or most likely a bit of both). If the latter, there is a further question concerning the extent to which they may be purposefully developed through social learning. On this, there exists an important literature that argues that they are attributes that should be purposefully reinforced as transboundary and intervention competences among environmental, including climate change, professionals (Pérez Salgado et al. 2012, 2014). Should we, however, expect *all* participating citizens to have them *all* of the time? Our view is that this is unrealistic. Also, it is not necessarily always desirable. If an aspect of communication of lived experience relates to how one feels, should not one be able to express emotions such as anger which might make others feel uncomfortable? The issue then becomes, therefore, one of how we manage the situation on the ground. Moreover, even if the attributes are developed purposefully and formally, how might we ensure that they are continually reinforced through actually practising them throughout our lives?

10.3.3 The Issue of Power

At the start of this chapter, we provided a neutral definition of institutions as the behaviours and practices that relate to the overall 'rules of the game', which are enforced formally through the law or informally through expected norms of behaviour that are derived from collective values. This definition, however, omits how these rules, norms of behaviour and collective values, come into being and how they are maintained and reinforced.

On this point, Mills (1959, 2000: 29) provides a useful contrasting definition of an institution in terms of its key actors and their roles in setting the underlying agendas. Thus: 'An institution is a set of roles graded in authority where the expectations of some men seem just a little more urgent than anyone else'. That is, 'some men' (*sic*) have more power and, by extension, it is these same 'men' who over time create the rules of the game, the expected norms of behaviour and the collective values that characterise an institution. Mills calls these characterisations 'symbols of justification' because of the way that they are used by people in the institution—the relatively powerless as well as the relatively powerful—to 'justify or oppose the arrangement of power and the positions within this arrangement of the powerful (Ibid. 36)… They become the basis for adherence to the structure of power or for opposing it' (Ibid. 40).

Thus power is everywhere institutionalised in our institutions through symbols. Institutionalising lived experience brings it into this system of graded authority where, some would argue, the purpose is to regulate it so that it does not burst out in unregulated and often unforeseen anger at policy decisions and interventions that it dislikes. Here it must further be stated that Mills is not thinking simply of power as coercion, but also as 'authority (power justified by the beliefs of the voluntarily obedient) and manipulation (power wielded unbeknown to the powerless)' (Ibid.).

Symbols of justification, therefore, are tools of both authority and manipulation. One issue that is raised by their presence concerns possible government *co-option* of lived experiential knowledge through institutionalisation which contrasts with our positive view as recorded in Sect. 10.2. There we presented institutionalisation of lived experience as a potential way of confronting power and privilege through expanding knowledge boundaries to create a social imagination. Co-option, however, normalises it. In other words, it is better for government to co-opt this non-scientific form of knowledge within the rules of the game so as to discipline it into the dominant way of thinking, than to leave it on the outside. In its most advanced form such disciplining takes place by manipulating the engagement space so that the citizens who bring knowledge of lived experience discover for themselves the 'truth' of the game as defined by the dominant discourse of the institution and its allies. Thus, Joseph Stiglitz once remarked while he was chief economist at the World Bank:

> George Bernard Shaw insightfully quipped: 'If you teach a man anything he will never learn it'.... Ortega y Gasset wisely suggested: 'He who wants to teach a truth should place us in the position to discover it ourselves'.... Thus if a global knowledge-based institution wants a country to learn a 'truth' about development, then it should help the local knowledge institutes and policy-makers... to learn it themselves — to make it a 'local social discovery'. (Stiglitz 1999: 211–212).

In this critical narrative, therefore, learning for oneself represents a normalisation of the same 'truth' for all actors operating within the institution. It further enables an appeal to the TINA ('There Is No Alternative') logic as a technique of government for purposes of persuasion and legitimation. Through a process of self-discovery and learning, we eventually arrive at the same universal truths. Although this process is often labelled 'depoliticization' in the critical literature, it is actually an effective political strategy of persuasion by which we are persuaded to persuade ourselves (Wilson 2007).

Related to co-option is the possibility that existing power might silence lived experience if it becomes institutionalised. Thus, Machin (2013: 80) refers to the inevitability of power relations in political discussion and quotes Iris Young: 'The norms of deliberation are culturally specific and often operate as forms of power that silence or devalue the speech of some people'. A few pages later she cites Chantal Mouffe on the eradicable dimension of power which blocks the possibility of giving all perspectives equal access to and manoeuvrability within the discussion (Ibid: 85, 86).

Similar conclusions to those reached by Machin, Young and Mouffe may be drawn from a rare empirical study of another environmental debate that took place during the first decade of this century, especially in affluent countries—that of genetically modified crops and food (GM). In the United Kingdom this took the form of a Government-promoted national public debate that also emphasised dialogue. Cook et al. (2004), however, found from interviews of 18 GM scientists, 15 non-experts and 10 outside advisers with diverse professional interests in GM that the scientists:

- Categorised everyone else as 'the public', to whom scientific method had to be explained, who had no expertise, and was superficial and irrational
- Used emotive language about GM opponents who they categorised by their sources and types of argument rather than by the arguments they were actually making

- Framed the discussion in terms of empirical objectivity, rather than trying to understand other frames
- Stuck rigidly to scientific language during communication, thereby excluding other dimensions of the debate
- Made 'simplification' the only concession to communicating with non-scientific audiences, and did not enter into the wider social and ethical arguments.

These findings resonate with the personal story of working with scientists in Chap. 3, Sect. 3.4. Cook et al. (2004) concluded of the GM debate: 'The debate is characterized by a heteroglossia of voices, some which have more authority than others. The evidence marshalled in this article provides a unique window into the ways in which scientists, key authoritative participants in the debate, deal with what they perceive to be an encroachment into their own discoursal terrain'.

This silencing claim contrasts with a more positive view that pervaded Sect. 10.2 above where public dialogue generates power. With respect to the latter, Norgaard (2011: 652) writes that philosophers such as Hannah Arendt and Jürgen Habermas describe the central importance of public dialogue to the generation of power and the project of democracy. By this she means that public dialogue, especially 'political talk', produces power through engaging the social imagination. She later quotes Arendt (Ibid: 816) directly: 'Power springs up for men when they act together' and powerlessness comes from being 'inattentively caught in the web of human relationships'. Thus dialogue and acting together generates power *with* others, but it also generates, through engaging the social imagination, the power *to* define what is worthy of public debate. Some years ago, Rowlands (1997) juxtaposed 'power with' and 'power to' against the way in which it is normally thought of as 'power over'. Arendt and Habermas hold a positive view which illustrates how 'power with' and 'power to' of previously unheard voices may challenge the conventional truth and power.

The fundamental, overarching issue of power, therefore concerns on one hand the extent to which institutionalising lived experience provides space for 'power to' and 'power with' through developing a social imagination, and on the other hand the extent to which it is co-option and silencing by the established powerful, manipulation towards a prevailing view of reality, and refusal of the powerful to engage on anything that lies outside of their own terms and frame.

Behind this latter somewhat gloomy view of the operation of power in actor engagements, especially those between scientists and non-experts, lies an assumption of positive feedback. This is the assumption that the exercise of power feeds the ability to exercise power in an ever-continuing cycle of mutual reinforcement. Through positive feedback, power consolidates elites. Traditionally these have been political elites and the owners of the means of production in the economy, but now we must include those who hold specialist knowledge such as climate scientists, and alliances between actors, especially alliances of actors who represent countries with perceived common interests.

Meanwhile, our own view on the power issue starts with the observation that it is everywhere in every human relation and cannot be ignored or wished away. Moreover, the exercise of power is crucial to making decisions. Without it there would be stalemate and no decision. As Mills (1959, 2000: 40) puts it: 'The problem of making or not making decisions is the basic problem of power'.

Given that, important for us when considering institutionalisation of lived experience in public policy making on climate change is how power is exercised and by whom. This involves examining the balance of power relations at any given time, the configurations of actors and the alliances and accommodations they make that allow moments of decision. Also important are the configurations of sources of power, including that of relevant knowledge. As we noted at the start of this chapter, the fact of recognising and institutionalising lived experience will accord it some knowledge–power. The notion of interdependence is also a relative source of power for citizens and their lived experience. Yet it is positive engagement of citizens and their lived experiences with other actors, especially science actors, on climate change that potentially creates a transformative alliance. While Cook et al's study of scientists and non-scientists in the GM debate is not encouraging about this possibility, there do exist climate scientists who wish to move forward and not be stuck intransigently in their own world views. See for example, the edited volume of Jaeger et al. (2012) whom we have cited several times already in this book. Nor are natural scientists the only people providing expertise. Engineers, for example, usually have to 'deal with' non-experts in their everyday work and so are possibly more attuned to working together with them (see 10.2.4 above).

Finally, to give the final word in Sect. 10.3 to the more positive view of power in relation to lived experience, it is to:

> Marshal knowledge of citizen lived experience, scientific and other expert knowledge to generate a transboundary social imagination on climate change that is able to set a public agenda on how to frame and intervene on the challenge.

10.4 Ways of Institutionalising

How might one institutionalise lived experience within policy- and decision-making processes on climate change? Again, in this section we claim no more than possibility and potential for extenuating positive outcomes from the spaces that are created. If, however, a full social imagination is too much to ask, at least we expect signs of a process that leads towards it.

Nowhere in the world as far as we are aware has there been any attempt at this kind of institutionalisation with respect to climate change policy making. Therefore we must search beyond, adapt and invent. We conduct our analysis within three broad categories of institutionalisation into policy making through: monitoring social media, government focus groups and independent environmental action groups. We then bring our exploration together at the end of the section (and the chapter) under the umbrella and concept of 'space'.

10.4.1 Institutionalisation Through Monitoring Independent Social Media

We referred to the possible use of social media for expressing lived experience and science engagement towards the end of Chap. 8. Hulme (2013: 258) refers to several Internet blogs that exist, some (but not all) of which are set up by climate scientists to engage, among others, the public. Box 10.1 describes five examples[2] that are mentioned by Hulme (Ibid.).

Box 10.1 A sample of climate blogs named by Hulme (Ibid.) All were active in February 2015

Bishop Hill is owned by Montfort (2014) a British climate sceptic. He describes himself as 'author, pamphleteer and one of the UK's best-best known critics of the so-called global warming consensus'.

Climate Audit is the blog of McIntyre (2014), a mining exploration company director. The blog examines in depth statistical climate data, questioning especially data that purport to show that twentieth century global temperature increase was unprecedented. While sceptically inclined on climate change, he tries to insist (not always successfully) that the blog contains reasoned discussion.

Climate Etc. is run by Curry (2014), an American climatologist and is intended as a forum for 'climate researchers, academics and technical experts from other fields, citizen scientists and the interested public to engage in a discussion on topics related to climate science and the science-policy interface'. She generally supports the global warming consensus of the majority of scientists and the United Nations Intergovernmental Panel on Climate Change (IPCC). She also supports the idea of climate change as a 'wicked problem' (Chap. 4, Sect. 4.2) and is critical of the IPCC for producing a frame that is too simplified to enable a broad understanding that includes social vulnerability. She calls for a rational discussion of climate change (Curry 2010).

Die Klimazwiebel (The Climate Union) is a blog run by German climate scientist von Storch (2014). He accepts the scientific evidence of global warming brought about by greenhouse gas emissions, but is concerned that climate scientists are at fault for often over-dramatising their science which over time reduces their credibility. Contributions are allowed in different languages, although practically this means German and English. The introductory paragraph states that 'the participants… are made of a diverse group of people interested in the climate issue; among them people who consider the man-made climate change explanation as true, and others who consider

[2]Obviously blogs may become defunct as well as be created and we have no way of knowing whether or not those shown in Box 10.1 will still be functioning when you read this.

this explanation false. We have scientists and lay people; natural scientists and social scientists, people with different cultural and professional backgrounds'.

Climate Dialogue (2014) is set up by the Dutch Ministry of Infrastructure and the Environment as a 'platform for discussions between (climate) scientists on important climate topics'. Discussion topics are selected by an editorial team which also selects the principal contributors, moderates the discussion and provides a summary at the end. The blog also states: 'The public (including other climate scientists) is also free to comment, but for practical reasons these comments will be shown separately'. Topics tend to be highly technical.

In the above list *Climate Dialogue* is the only (Dutch) government-sponsored blog. All except this blog (which has selected, invited contributors to its main discussion) and the sceptical *Bishop Hill* blog have posting rules that attempt to warn off those not interested in serious dialogue and debate. They do not in general want shouting, obscenities, insults or unsubstantiated statements. Some discussions evoke hundreds of responses. For example, Judith Curry started a *Climate Etc.* discussion on 19 November 2014, titled: 'Can cows help save the planet: some interesting new ideas about the role of soils, farming and livestock in fighting global warming'. Three days later, when we accessed the discussion, it had received 494 comments.

There are more blogs devoted to climate change than those described in Box 10.1, and even more climate scientists who contribute to other, more general, blogs. Rather than set up its own site, any Government could with relative ease employ a small team to monitor and analyse the postings on them (although it might want to restrict itself to those that emanate from the country it governs).

There are issues of course, particularly around the representativeness of the blogs and the general 'low' quality of the data that they provide. While some postings are themselves substantial and reasoned, most are too short—labelled 'snacking' by Peter Bazalgette (quoted in Ford 2014: 20) who is chairman of Arts Council England and an investor in social media. Some seem intent on discrediting whoever they do not like (posts on *Bishop Hill* when we accessed the blog in November 2014 were prone to that).

Newspaper columnist Tett (2014: 54) comments that social media platforms such as Twitter 'are an effective way to create a loud political noise but not a good place to create thoughtful debate. Attention spans are so short that everyone needs to shout to be heard. Extremism tends to dominate'. The general problem appears, as we suggested in Chap. 8, to be that digital literacy is in its infancy even among those in the affluent world who have been there at the start. The sheer quantity of posts, however, can possibly make up for low quality if they are mined carefully, although the extent to which they will ever express lived experience in its full richness—deliberately or by chance—remains open to debate.

10.4.2 Direct Governmental Institutionalisation Through Focus Groups

Any government in theory could establish forums where it listens to a range of lived experience perspectives across the social categories that fall within its province—rich/poor, women/men, different ethnic and religious entities for example. In reality listening forums would most likely be set up in a representational democracy such as exists in many parts of the world, where the Government is elected and enjoys a degree of popular legitimacy. Within them, Government would take note of the different lived experience perspectives and also, as it does already, continue to hear the scientific perspectives. Then it and it alone would come to an overall judgement and make its policy decision.

Government could also do a limited exercise of listening to lived experiences purely through 1:1 interviews with a representative sample of people, and subsequently piece everything together. Generating discussion among a group of participants, however, offers further opportunities, where the multiple interactions potentially generate insights and depth that are difficult to achieve with 1:1 situations.

One type of national forum might be a social media site as examined in Sect. 10.4.1, but where government establishes its own blog rather than monitor those that exist independently of it. Unlike the Dutch Government-sponsored *Climate Dialogue* (Box 10.1), such a blog would be deliberately aimed at capturing a range of citizen lived experiences of climate change, and/or citizen-science dialogue. Even here, however, it would be difficult to overcome problems of representativeness and the low data quality that we mentioned in Sect. 10.4.1, unless it was open only to participation by a relatively small number of selected, invited citizens/scientists.

Face-to-face focus groups are another way of government approaching institutionalisation of lived experience with relatively small numbers of invited participants. These have a long history, a reference of early use being in World War 2 to test effectiveness of propaganda messages. Basically, they comprise 6–12 people who are invited to discuss a specified topic—for example what climate change means to them. Sessions are usually moderated and observed. Representativeness need not pose a major challenge as government could (which is not to say that it necessarily 'would') invite a cross section of society, including climate scientists, to participate depending on the purpose. As a form of qualitative research it would face the usual challenges of data analysis (Chap. 4), but could generate very useful insights of lived experience as well as stimulating deeper conversations that include the sciences.

The UK Government (2014), for example, suggests combining focus groups with 1:1 interviews. Both are described as 'unstructured' although this is doubtful as the topic is provided in advance, while a trained moderator keeps the discussion 'focussed' and 'on topic'. Sessions normally last about 12 h, and communication materials and interactive techniques, such as role plays, may form part of the exercise. The UK Government adds that focus groups are 'useful techniques for exploring and mapping reasons for attitudes and behaviour, understanding how target audiences approach issues or may be encouraged to change'. Thus, they may be used to tap into lived experiences of what the Government describes as 'the general population or hard to reach groups' (Ibid.).

One possibility is to use social media to complement focus groups, that is, use the quantity of data from social media to identify key topics for substantive discussion in focus groups. Also note that, while focus groups are usually thought of as involving face-to-face interaction, they could themselves be conducted by digital means.

So far when considering direct government institutionalisation of lived experience we have examined the national picture. Face-to-face focus groups, complemented by social media, could of course operate at the municipal level and focus on more local issues of lived experience in relation to climate change impacts and interventions. Moser and Dilling's (2007) edited collection contains several chapters about climate change actions that have involved municipal/state authorities and citizens in the United States. Thus, in some localities it is quite possible to graft lived experience onto existing governmental structures.

Many municipalities, moreover, have twinning arrangements with other towns and cities across the globe, potentially providing an avenue for international dialogue on climate change. In Chap. 8 (and also briefly above in this chapter) we reported on practitioner-to-practitioner partnerships between the UK and Ugandan municipalities, involving environmental health engineers. So too might there be citizen-to-citizen partnerships who share experiences of climate change, thus providing the possibility of contributing towards international awareness. One of the Moser and Dilling contributors, Young (2007), describes the work of the worldwide 'Cities for Climate Protection program' which was established in 1993 by the International Council for Local Environment Initiatives (ICLEI), itself partly a product of the United Nations Environment Programme. Its main focus is reducing urban greenhouse gas emissions among the participating cities, but the point for us is that structures such as this ('Cities for Climate Protection' is not the only international network of its kind) exist potentially for sharing lived experiences across geographical divides.

10.4.3 Independent Institutionalisation Through Environmental Action Groups

The social media sites to which we referred in Sect. 10.4.1 and the examples we gave in Box 10.1 would be examples of independent institutionalisation if Government were to monitor them systematically for what they reveal about lived experiences. They are not, however, purposeful in the sense of providing direct institutionalisation of lived experience into climate change policy making. Nevertheless, they do form part of public action in policy making in the sense explored in Chap. 9. Something from the discussion in these sites will filter through to Government even without monitoring them, and the issues they raise will not just be confined to those sites.

In Sect. 10.4.3, however, we are more concerned with *purposeful* institutionalisation of lived experience into climate change policy making through forums that include but are independent of government. For this, we focus on the concept of Environmental Action Groups (EAGs) that we introduced briefly towards the end of Chap. 8. Being based in a poor country—Zimbabwe—there is much that the rest of the world can learn from EAGs in terms of stakeholder engagement across boundaries and institutionalisation of lived experience into policy making and intervention.

Conceived and established by the regional NGO Environment Africa, EAGs are multi-stakeholder forums for considering environmental issues and intervention in a locality.[3] For example, Johnson and Wilson (2000, 2009b) report their action research project with the EAG for the Zimbabwean mining town of Bindura. The main stakeholders comprised Environment Africa, the local municipal authority environmental health department, local offices of relevant national ministries, the mines, local retailers and voluntary groups in the town. In Johnson and Wilson's action research, the last included a Christian church-based group and an association of widows of AIDs victims.

Environment Africa has established several such EAGs over Southern Africa. The specific composition varies depending on the local context, but will typically comprise governmental, private sector and civil society actors, where the inclusion of:

- *Local civil society representatives* legitimises their experiential knowledge which also enables them to make links to other concerns that are possibly more important to them. Thus, the Widow's Association in Wilson and Johnson's action research was not primarily concerned with 'the environment', but about livelihood opportunity, in this case collecting the recyclable components of waste collection and being paid for their sale or reuse (Ibid).
- *Government* (local or national, or both as in Johnson and Wilson's action research) provides an important perspective to the discussions that will be different from those of the private sector and civil society groups. Moreover, its very participation *legitimises* the EAG, while through its participation it is in turn *legitimised by* the EAG.
- *Locally based private sector representatives* brings yet a further dimension to the space provided by the forum. An unpublished part of Johnson and Wilson's action research (Ibid) suggests that private sector representatives do not simply express their interests, but continually raise issues of practicality, especially around interventions. Thus, they ask questions about the finances, the management, who runs the show and efficiency. While it is wrong to say that other groups ignore these matters, their whole discourse is not structured around it as it is with private sector representatives.

The Southern African EAGs would probably not exist, however, without the championing and coordination by Environment Africa. The NGO has acted as a classic boundary organisation as mentioned in Chap. 8, that mediates between

[3]Recently the concept of EAGs as conceived by Environment Africa has evolved into 'Private, Public, Community Partnerships' (PPCPs). The substance of what we write here about EAGs also applies to PPCPs.

the different social worlds of the stakeholders (Ng'ang'a 2013: 2). In its original formulation by Guston (cited in Ng'ang'a 2013: 57), boundary organisations exist at the interfaces of scientific and other communities. Clearly, the EAGs in Southern Africa are conceived more broadly than this, as being at the interfaces of a spectrum of stakeholders with environmental interests and concerns. Guston's formulation, however, does also provide a hook in terms of one of the main story lines of this book—the boundaries between scientific and lived experience forms of knowledge.

Thus, as is typical of boundary organisations (Guston 2001), Southern African EAGs:

- Convene stakeholder engagement across their different social worlds
- Translate stakeholder discourses and try to arrive at a common discourse (or language)
- Inspire collaboration between the stakeholders
- Mediate conflicts.

In a similar vein, much of the EAG work is done through boundary agents who are employed by Environment Africa with the necessary skills and abilities to 'straddle the multiple and varied boundaries between stakeholders' and to 'develop trust and mutual respect across social worlds despite power asymmetries' (Ng'ang'a 2013: 57–8). EAG meetings provide the spaces for these processes and thus constitute what the theory calls 'boundary objects'.

Because they involve government in some form, EAGs as conceived by Environment Africa do appear, therefore, to offer the possibility of institutionalising lived experience in public policy and action-making. They have some similarity with the 'new hybrid institutions' that Corner and Groves (2014; see also Chap. 7 Sect. 7.3) advocate for negotiating the science and public engagement interface, but are more piecemeal and do not suggest any claims to institutional grandeur. Key questions, however, concern the adaptation of EAGs to issues of climate change and their replicability as a model to be adopted across different countries.

It would be a mistake to try and replicate EAGs exactly across different country contexts. One size cannot possibly fit all in a diverse world. The key common element is that they do act as boundary organisations to bring together stakeholders and their diverse social worlds, and that they at least are recognised by governments as legitimate players. Beyond that, the EAG model as formulated by Environment Africa may be adapted freely. For example, in some contexts it might be appropriate to focus them as 'Climate Change Action Groups' (CCAGs) while in others, climate change will be considered within the larger concept of EAGs. One reason why we might prefer the latter is that it places climate change within a broader remit, allowing for consideration alongside other topics and issues. As already stated, in the Southern African EAGs, the diverse stakeholders who form their core groups broadly comprise relevant arms of government, the local private sector, and civil society organisations. When considering climate change, it would be important to add scientists to the core group and to specify more specifically the range of civil society actors. For example, it would be possible to specify that they cover different social, including ethnic and religious, groups.

In the United Kingdom we have referred in Sect. 10.2.1 above to the 'Institute for Government' whose remit is to work to increase government effectiveness. It is essentially a 'think tank' that works across the main political parties and advises government. This would not be a bad remit for a network of EAGs in the UK. While close to Government, the Institute for Government is, like Environment Africa in Zimbabwe, a registered non-profit organisation and independent of it. An EAG network could advise local and national government. It would not be a 'think tank' in the usual sense and certainly would not attempt to provide the latest climate science. Rather, it would represent a unique synthesis of diverse stakeholder perspectives, including the lived experiences of citizen groups.

The experience of EAGs in Southern Africa suggests that a strong coordinating championing body is essential that acts as a boundary organisation as discussed above. A registered charity (which could be an arm of a separate environmental NGO) is the obvious choice, without excluding the possibility of different coordination mechanisms in different contexts. It would adopt values and behaviours as highlighted in Sect. 10.3.2, such as listening, being respectful to all and reflective. Developing these values and behaviours as transboundary and intervention competences among its individual boundary agents, as described by Pérez et al. (2012, 2014—see also above) would be an important facet. One function of the boundary agents then would be to diffuse the same values and behaviours among the diverse stakeholders of the EAGs, while also being able to manage situations where they will not be evident.

As well as advising Government, an EAG could have specific remits, for example to:

- Raise awareness through special events to which the wider community is invited
- Conduct stakeholder analysis of local interventions that are planned on climate change. A stakeholder analysis explores the likely negative and positive impacts—both immediate and short-term, and long-term for future generations—of the proposals on different stakeholders and issues associated with possible opposition. With many local interventions, the range of stakeholders will be larger than that represented by the EAG's core members. It should, therefore, have the powers to co-opt and/or consult others for particular issues, for example a residents' group that may be affected by a proposed intervention. Also any local climate change intervention will have some kind of knock-on effect the far side of the world however insignificant it feels. Thus stakeholders that are considered should include those who live very different lives in other countries, as well as those locally, even if they cannot be co-opted directly onto the EAG.
- With respect to the last point, a network of EAGs internationally might be established, interacting primarily through social media, to help widen the horizons of members.

Of course there are issues with establishing an EAG or similar in any country. These issues include the consent of Government; financing and material resources for activities; whether a critical population mass is needed (which could restrict them to towns and cities); selection of core members; and incentives to give up time and energy to participate.

On the last of these issues, we suggested in Sect. 10.2 reasons why different stakeholders might wish to have lived experiential knowledge institutionalised, and hence incentives for establishing and maintaining an EAG. Beyond that, a general incentive for all stakeholders would be that they obtain the ear of government.

10.4.4 Bringing It Together: Institutionalising as Space Making

Thus far in Sect. 10.4 we have considered the possible mechanisms for institutionalising lived experience within Government policy making and intervention on issues related to climate change. Apart from Government trawling the relevant social media that already exist on the subject, both Government-sponsored focus groups and independent EAGs (or CCAGs) provide deliberately constructed, 'invited spaces' for engagement between diverse actors (Cornwall 2004).

We stress again, however, that we have only outlined possibility and potential in Sect. 10.4. Nowhere do we claim to have overcome power differentials that represent different social positions of groups, although in Sect. 10.3 we argued that power may be positive as well as negative, and that there is more than one source of power at play. We also reiterate what we stated in Sect. 10.3, which is that any space will have its own dynamic and circulation of power and that it is impossible to predict what will happen. As Cornwall (Ibid. 80–81) states: 'Spaces created with one purpose in mind may be used by those who engage with them for something quite different'. Thus, while a broad aim might be to bring different stakeholders together in dialogue and advise government, it is possible for the space created to develop deeper transboundary dialogues and debates. It may even start to engender a social imagination and construct interdependency.

Of course, spaces might not do these things or only show signs of doing them in a random, even fickle, way. It may also appear that simple co-option by government and other powerful interests has taken place. However, just as the saying goes, 'It takes two to tango', it also takes two to allow co-option and there is always implicit negotiation between co-opter and co-opted. What we hope for is not necessarily perfection in terms of our interests as authors of this book—which is developing a shared, transboundary, social imagination—but something that suggests a process, however imperfect, along those lines.

In this, we remind ourselves of another old adage—perfection is the enemy of the good. For us, this means that it is bad faith to attempt nothing simply because you believe that it will never reach the perfection that may only be achieved through first overcoming all of the structural barriers to transformative change.

References

Abbott, D., & Wilson, G. (2014). Climate change: Lived experience, policy and public action. *International Journal of Climate Change Strategies and Management, 6*(1), 5–18.

Adamson, P. (2013). A measure of progress. *New Internationalist, 460*, 12–15.

Bynum, W., & Porter, R. (Eds.). (2005). *Oxford dictionary of scientific quotations*. Oxford: Oxford University Press.

Climate Dialogue. (2014). Blog of the Dutch Ministry of Infrastructure and the Environment. Accessed November 22, 2014 www.climatedialogue.org.

Chambers, R. (1997). *Whose reality counts: putting the first last*. London: ITDG Publishing.

Cooke, B., & Kothari, U. (2001). *Participation: The new tyranny?*. London: Zed Books.

Cook, G., Pieri, E., & Robbins, P. (2004). 'The scientists think and the public feels': Expert perceptions of the discourse of GM food. *Discourse and Society, 15*(4), 433–449.

Corner, A. & Groves, C. (2014, September 4). Breaking the climate change communication deadlock. In *Nature Climate Change* (pp. 743–745).

Cornwall, A. (2002). *Making spaces, changing places: Situating participation in development.* Working paper 170, Institute of Development Studies, University of Sussex, Brighton.

Cornwall, A. (2004). Spaces for transformation? Reflections on issues of power and difference in participation in development. In S. Hickey, & G. Mohan (Eds.), *Participation. From tyranny to transformation*? London: Zed Books.

Craig, D., & Porter, D. (2003). Poverty reduction strategy papers: A new convergence. *World Development, 31*(1), 53–69.

Curry, J. (2010). Statement to the sub-committee on energy and environment of the United States House of Representatives. Accessed November, 22 2014 judithcurry.com.

Curry, J. (2014). Climate etc. blog of Judith Curry. Accessed November, 22 2014 judithcurry.com/about/.

Ford, J. (2014). YouTube: Taking a bite out of TV'. In *Financial Times* (15/16 November). London: Pearson Publishing.

Foucault, Michel. (1979). Governmentality. *Ideology and Consciousness, 6*, 5–22.

Foucault, M. (1980). Truth and power. In C. Gordon (Ed.), *Michel Foucault: Power/knowledge: Selected interviews and other writings 1972–1977*. London: The Harvester Press, Pearson Education Ltd.

Gawande, A. (2014, November 15/16). The science interview: 'We lose sight of what people might be alive for'. In *Financial Times*. London: Pearson Publishing.

Guston, D. H. (2001). Boundary organisations in environmental policy and science: An introduction. *Science, Technology and Human Values, 26*(4), 399–408.

Hallsworth, M., Parker, S., & Rutter, J. (2011). *Policy making in the real world: Evidence and analysis*, Institute for Government. Accessed November, 17 2014 available at: http://www.instituteforgovernment.org.uk/sites/default/files/publications/Policy%20making%20in%20the%20real%20world.pdf.

Hulme, M. (2009). *Why we disagree about climate change*. Cambridge: Cambridge University Press.

Hulme, M. (2010a). The year climate science was redefined. In *Exploring climate change in science and society: An anthology of Mike Hulme's writings, interviews and speeches*. London: Earthscan/Routledge.

Hulme, M. (2010b). Moving beyond climate change. In *Exploring climate change in science and society: An anthology of Mike Hulme's writings, interviews and speeches*. London: Earthscan/Routledge.

Hulme, M. (2013). *After Climategate... Never The Same. In Exploring climate change in science and society: An anthology of Mike Hulme's writings, interviews and speeches*. London: Earthscan/Routledge.

Isaacs, W. (1993). Taking flight: Dialogue. *Collective Thinking and Organizational Learning, Organizational Dynamics, 22*, 24–39.

Jaeger, C., Hasselmann, K., Leipold, G., Mangalagiu, D., & Tabara, J. D. (Eds.). (2012). *Reframing the problem of climate change: From zero sum game to win-win solutions*. London: Earthscan/Routledge.

Johnson, H., & Wilson, G. (2000). Biting the bullet: Civil society, social learning and the transformation of local governance. *World Development, 28*(11), 1891–1906.

Johnson, H., & Wilson, G. (2007). North–South/South–North partnerships: Closing the mutuality gap. *Public Administration and Development, 26*(1), 71–80.

Johnson, H., & Wilson, G. (2009a). Learning and mutuality in municipal partnerships and beyond: A focus on northern partners. *Habitat International, 33*(2), 210–217.

Johnson, H., & Wilson, G. (2009b). *Learning for development*. London and New York: Zed Books.

Kolb, D. (1984). *Experiential learning: Experience as the source of learning and development*. Englewood Cliffs, New Jersey: Prentice Hall.

Leroi, A. M. (2014). *The lagoon: How Aristotle invented science*. London: Bloomsbury Circus.

Machin, A. (2013). *Negotiating climate change: Radical democracy and the illusion of consensus*. London: Zed Books.

McIntyre, S. (2014). Climate Audit blog of Steve McIntyre. Accessed November, 22 2014 Climateaudit.org.

Mills, C. W. (1959, 2000). *The sociological imagination*. Oxford: Oxford University Press.

Montfort, A. (2014). *Bishop Hill blog of Andrew Montfort. Click on 'the dissentient' to access the blog*. Accessed November, 22 2014 Bishophill.squarespace.com.

Moser, Susanne, & Dilling, L. (Eds.). (2007). *Creating a climate for change: Communicating climate change and facilitating social change*. Cambridge: Cambridge University Press.

Ng'ang'a, E. (2013). *A boundary organisation for the south: The success and challenges of TWAS, the academy of sciences for the developing world, in attaining power for scientists of the south*. Ph D thesis, The Open University, Milton Keynes.

Norgaard, Kari M. (2011). *Living in denial: Climate change, emotions and everyday life*. Cambridge, Massachusetts: The MIT Press.

Ostrom, Elinor. (1990). *Governing the commons: The evolution of institutions for collective action*. Cambridge: Cambridge University Press.

Pérez Salgado, F., Wilson, G., Van der Klink, M (2014). Transforming academic knowledge and the concept of lived experience: Intervention competence in an international e-learning programme. In S. Caeiro, W. Leal Filho, U. M. Azeiteiro (Eds.), *'E-learning and sustainability' in the series 'Environmental Education, Communication and* Sustainability' (pp. 59–69). Pieterlen: Peter Lang.

Pérez Salgado, F., de Kraker, J., Boon, J., & Van der Klink, M. (2012). Competences for climate change education in a virtual mobility setting. *International Journal of Innovation and Sustainable Development, 6*(1), 53–65.

Polanyi, Michael. (1958). *Personal knowledge*. Chicago: University of Chicago Press.

Rakodi, C. (2000). *Technical decisions or interactive processes? Towards collaborative planning and decision making in towns and cities*. Paper presented to 'Urban Futures 2000', University of Witwatersrand, Johannesburg (10–14 July).

Regan, K. (2007). A role for dialogue in communication about climate change. In S. Moser, & L. Dilling (Eds.), *Creating a climate for change: Communicating climate change and facilitating social change*. Cambridge: Cambridge University Press.

Rowlands, J. (1997). *Questioning empowerment: Working with women in Honduras*. Oxford: Oxfam GB.

Smith, J., Clark, N., & Yusoff, K. (2007). Interdependence. *Geography Compass, 1*(3), 340–359. doi:10.1111/j.1749-8198.2007.00015.x.

Stern, Nicholas. (2010). *Blueprint for a safer planet: How we can save the world and create prosperity*. London: Vintage Books.

Stiglitz, J. (1999). Scan globally, reinvent locally: Knowledge infrastructure and the localization of knowledge. Keynote address to the first Global Development Network Conference, Berlin (December). In H. J. Chang (Ed.), *Joseph Stiglitz: The rebel within*. London and New York: Anthem Press. Accessed November, 21 2014 via: http://books.google.co.uk/books?id=XoIIfHk7eQUC&pg=PA194&lpg=PA194&dq=Joseph+Stiglitz+Scan+globally&source=bl&ots=HavucS46mg&sig=WTU8_BylDHZhFvtbP8XQd9tYcNQ&hl=en&sa=X&ei=NDhvVK_GMIPDPJv2gJAG&ved=0CDMQ6AEwAw#v=onepage&q=Joseph%20Stiglitz%20Scan%20globally&f=false.

Tett, Gillian. (2014). *Financial Times (15/16 November)*. London: Pearson Publishing.

Thomas, A. (2009). Local governance and sustainable development in Wales: Making a difference. In G. Wilson, P. Furniss, & R. Kimbowa (Eds.), *Environment, development and sustainability: Perspectives and cases from around the world*. Oxford: Oxford University Press.

UK Government. (2014). *Focus groups, mini-groups and 1:1 interviews*. Accessed November, 22 2014 https://www.gov.uk/service-manual/user-centred-design/user-research/focus-groups-mini-groups-interviews.html.

von Storch, H. (2014). *Die Klimazwiebel blog of Hans van Storch*. Accessed November, 22 2014 http://klimazwiebel.blogspot.co.uk/.

Karl, W. (1995). *Sensemaking in organizations*. London, Thousand Oaks, New Delhi: Sage Publications.

Wilson, Gordon. (2006). Beyond the technocrat? The professional expert in development practice. *Development and Change, 37*(3), 501–523.

Wilson, Gordon. (2007). Beyond the dichotomies in technology for development. *Geography Compass, 1*, 119–135. doi:10.1111/j.1749-8198.2007.00005.x.

Wilson, Gordon. (2008). Our knowledge ourselves: Engineers (re)thinking technology in development. *Journal of International Development, 20*, 739–750.

Wilson, Gordon, & Johnson, H. (2007). Knowledge, learning and practice in North–South municipal partnerships. *Local Government Studies, 33*(2), 253–269.

Young, A. (2007). Forming networks, enabling leaders, financing action: The cities for climate protection campaign. In S. Moser, & L. Dilling (Eds.), *Creating a climate for change: Communicating climate change and facilitating social change*. Cambridge: Cambridge University Press.

Chapter 11
A Public Action Approach to Knowledge and Intervention to Meet the Climate Challenge

Abstract Lived experiences and science provide different but partial realities of climate change. Engaging lived experiences, however, does more than add another form of knowledge to the phenomenon. It adds further to our understanding of how people, as citizens, respond to the science, policy proposals and planned interventions. Lived experiences, moreover, are extremely diverse. This potentially allows for knowledge enrichment through engagement, but all too often it results in power dynamics between actors where the most powerful—which include national governments, international institutions and the corporate sector—do their utmost to define the truth of a situation and exclude much of the diversity of knowledge that actually exists. A public action approach to knowledge inherently recognises the diversity that arises from lived experiences and the power dynamics. It further illustrates how, in an interdependent world, no group's power is absolute and even the most vulnerable are able to lever a place on the climate change agenda at all scales, through instrumental sharing for mutual gain and political legitimacy agendas, alliances with others, NGOs and value-based social movements. We recognise, however, the unruly, messy process of knowledge construction that makes public action what it is. There are possibilities for institutionalising it to some extent and making it less unruly, but they depend on policy makers engaging in a process that shares, rather than simply extracts, knowledge, and on engaging lived experiences from diverse social groups. The potential prize for productive engagement is a collective transboundary social imagination that enables citizens to establish the public agenda surrounding climate change.

11.1 Introduction

The early chapters of this book contrasted lived experiential knowledge and knowledge provided by the sciences. Underlying the contrast was our argument that the two forms of knowledge might be very different, but they each could only be partial representations of reality with their own (partial) validities. They could, however, become potentially enriching complements to each other.

D. Abbott and G. Wilson, *The Lived Experience of Climate Change*,
DOI 10.1007/978-3-319-17945-2_11

This argument was carried through in later chapters, and indeed forms the basis for promoting engagement between lived experience and the sciences to create a broader and deeper knowledge about climate change. These later chapters, however, have also taken us beyond a simple complement towards understanding how we—as inhabitants of this Earth—respond to scientific and other expert knowledge about climate change and to the interventions that are initiated to address the challenge.

This is because lived experience is more than just a missing link that provides knowledge of the ways in which climate-linked events are experienced at local levels. It certainly fulfils that function but simultaneously it is a filter of scientific knowledge and policy communications. Put another way, we interpret scientific knowledge and policy communications through our lived experiences and respond positively or negatively, with a spectrum of shades in between.

In this way, lived experience provides a coherent account about how and why people respond in the ways they do. The response might be in the form of personal actions that people—as citizens and consumers—are prepared to take to, say, lower their personal carbon footprints (Chap. 6). It might go further and move towards activism with the general aim of combating climate change and related challenges (Chap. 1). A contrasting response, however, might be in the form of overt community opposition to mitigation interventions which are understood to have adverse impacts locally (Chaps. 1 and 6), or more passive socially organised denial as articulated by Norgaard (2011) in her anthropological study of a Norwegian community (Chap. 7). In poor developing countries, the response in the form of adaptation to climate change by individuals and groups may be less a matter of choice and more a matter of survival.

The diversity of responses matches the diversity of lived experiences, where dimensions of lived experience are shared with others. Diversity of lived experience has been a key motif that runs throughout the book. It derives partly from the socio-economic contexts in which lived experiences are embedded. A poor rural person living in a developing country will have a very different lived experience from a middle class person living in a rich country. The differences that are associated with gender, ethnicity and religion provide further contextual influences that generate diverse lived experiences. Against this backdrop climate-related events whose impacts may or may not be shared with others, and the reflections—individual and collective—on the actions that we take in response to such events add further diversity. The diversity of lived experiences are also a product of the people we meet and engage with through life, including most importantly close family members, and the information and knowledge we receive from a variety of sources.

The paragraph above has restated the conceptualisation of lived experience that we developed in Fig. 1.2 of Chap. 1. As described here, diversity is inherent to the conceptualisation, further complicated by the fact that many of the influences such as climate-related events, actions and reflection and engagements with others are dynamic, meaning that any individual lived experience must also be so.

In our view, we should celebrate this diversity. Why? Because through engagement with one another, diversity helps generate new knowledge of how to do things differently. People with similar mindsets, perspectives, experiences don't create much new knowledge—they tend to focus on doing better what they already know about. At its most optimistic, engagement with knowledge, diversity creates a transboundary social imagination—that quality of mind that helps us to use information and to develop reason in order to achieve lucid summations of what is going on in the world and of what may be happening within ourselves (Mills 1959/2000: 5; see also Chaps. 2 and 8).

Alas, diversity in the real world does not work like that. In social terms it all too often translates into inequality and power differences between individuals, social groups, scientists, private, public and NGO sectors, and nation states, where the most powerful do their utmost to define the truth of a situation and exclude much of the diversity of knowledge that actually exists.

These considerations return us to how policy around climate change, at all scales, is developed from a process of public action (Chaps. 9 and 10) whose two principal defining features are diversity of interests and power relations between actors in influencing outcomes. The rest of this final chapter considers the public action approach specifically in relation to knowledge, also defined by its diversity of forms and where some forms are more powerful and count for more in climate change policy debates than others.

A public action approach to knowledge as espoused in Sects. 11.2 and 11.3 unifies the concerns of our book.

11.2 A Public Action Approach to Knowledge

We start by casting in terms of power relations, cameos from two cases that have appeared in previous chapters, plus a third that is new to this book.

Cameo 1: levering power through knowledge alliances

Chapters 1 and 5 reported the experiences of the Ogiek and Batwa forest communities in East Africa, where climate change mitigation policies to conserve forests exacerbate existing vulnerabilities. Powerless by themselves to make more than minor inroads into national, and still less international, policies on forest preservation, these communities have allied with similar groups in other countries, such as the international human rights NGO, The Minority Rights Group.

This kind of public action to ally with minority groups elsewhere is first and foremost a strategy of levering greater power—to force their exacerbated vulnerability onto international climate change policy agendas so that it becomes known. Such leverage is also observed in the Alliance of Small Island States (AOSIS) which, with over 40 member countries, has brought knowledge of their particular vulnerability to rising sea levels to the international climate change agenda.

Cameo 2: the arrogance of (some) climate scientists

At the end of Chap. 3, one of us (the authors) reported a personal experience of working with physical scientists after being asked to lead a team to develop a new University Masters Programme on climate change:

> The team consisted of colleagues from my department who have known me for a very long time, but some resentment was apparent in that as I am a social scientist, and a senior academic (Professor), the 'scientists' felt they had superior knowledge of climate change and that I lacked authority both in my senior rank and as a social scientist. What followed

was massive disruption, caused by one colleague in particular who felt threatened by the process and led other newer and younger members of staff into doing the same. In the end, the idea was abandoned causing much stress and ill feeling within the department.

This cameo reveals the arrogance of a significant number of climate scientists in refusing to recognise the partiality of their knowledge, a refusal which leads directly to the exclusion of other (in this case qualitative social science) knowledge. It reflects the power of the physical sciences in terms of setting the climate change agenda as still reflected to an extent in the IPCC Assessment Reports. See also Chap. 10, Sect. 10.3.3, where we cite a study showing that scientists exhibited similar behaviour during the public debate in the United Kingdom about genetic modification.

Given that there was no public agenda on climate change before the physical sciences came to grips with the topic, which took well over a century to mature, we owe a large debt to them for establishing one over the years and for reinforcing it with robust analysis. This, however, is no reason for trying to maintain a closed agenda that excludes other forms of knowledge.

We should note that this personal experience from Chap. 3 also reports more enlightened physical scientists who acknowledge the need for a broad knowledge agenda, so the situation is dynamic, perhaps even moving in a better direction. Nevertheless there remain sufficient numbers who allow the power of their science to turn into arrogance.

Cameo 3: the corporate energy sector presence at recent United Nations climate change conferences

The corporate energy sector has few problems installing itself onto United Nations international climate change deliberations. Indeed, the 2013 United Nations 19th Conference of the Parties (COP—the annual gathering of governments since 1995 that are party to the United Nations Framework Convention on Climate Change) in Warsaw was heavily criticised for its large (overwhelming, according to some observers) influence. Then, at the following COP20 in Lima towards the end of 2014, the Global Carbon Capture and Storage (CCS) Institute hosted an event, 'Why Divest from Fossil Fuels When a Future with Low Emissions Fossil Fuel Energy Use is Already a Reality?' Its panel of experts included David Hone, chief climate adviser for the international energy giant Shell, and Milton Catelin, Chief Executive Officer of the World Coal Association. The aim of the event: to save the threatened fossil fuel industry through promoting knowledge claims about CCS—the use of technologies to capture waste carbon dioxide emissions and then store them underground—as an alternative to reducing their role in energy generation.

The above cameos illustrate how knowledge-power is embedded in social institutions (Chaps. 2 and 10). Science, with its shared 'rules of the game' concerning empirical experimentation and robust findings may be regarded as one such institution. In stark contrast, the Batwa and Ogiek communities of East Africa, and their context-bound knowledge of what climate change mitigation policies of forest preservation mean for them, are excluded to a large extent from social institutions (for example, that of inter-marriage with people from more dominant communities) and formal institutions of the state in their respective countries. They respond by being part of an international alliance.

The attempts of forest communities to force their way on to international agendas are also a far cry from the corporate sector which represents institutional rules of the game concerning capitalist processes of free market competition and innovation, enabled, but not coordinated, by states or supra-national political entities. With respect to climate change negotiation processes, the corporate energy sector, especially, feels under threat as most climate mitigation policies are designed to reduce carbon dioxide emissions into the atmosphere, emissions which arise primarily from the combustion of fossil fuels to release useful energy for heating, transport and industry. The sector, however, is able to draw on significant financia resources with which to make its voice heard, while its economic arguments resonate with almost every state on the planet. We may also be sure that should there ever be an international agreement to phase out fossil fuels, the corporate energy sector will have the resources to adapt to the new conteFor many who are concerned about climate change, institutionalised power makes them feel despondent that anything meaningful can be done. There are ways, however, of countering such power which have roots in our notion of human interdependence, graphically represented by images of the earth as having no internal boundaries, only an outer one demarcated by outer space. As one person who we asked to describe his lived experience of climate change stated: 'In the 1960s the colour photographs of the world from outer space taken by National American Space Agency (NASA) emphasised both the beauty of the earth and its fragility, and were some of the first images which did not come tagged with some national boundary or political jurisdiction encoded in the image' (Abbott 2015). Interdependence may also be accommodated within the frame of 'climate change as a threat and war' (Chap. 7) because, unless everybody is on board, we are all doomed.

xt—for example, simply by moving into renewable energy and insulation (Chivers 2015).

Overall, therefore, our interdependence means that exclusion from the climate change debate is not an option and for this reason it becomes a source of power for the weak and vulnerable of the earth. From here, there are two broad ways of exercising that power: instrumental and value promotion.

By instrumental we mean using our interdependence to marshal useful knowledge about climate change and intervention to meet its challenge. Interdependence stakes a claim in our consciousness because climate change respects no borders and will affect everybody eventually. It is therefore in our private interest to have a broader utility interest in the responses of others to it.

Thus, we might wish, for example, to set aside our internally constructed, earthly boundaries to examine closely and share the knowledge—often local or indigenous—that is embedded in community-based adaptation interventions (Chaps. 5 and 6) to see if it is useful to us. We might take seriously the title and message of Stern's (2010) book, *Blueprint for a safer planet: how we can save the world and create prosperity.* We might try to improve the public understanding of the science of climate change, to bring everyone round to one way of thinking.

In practice, however, there is no such thing as purely instrumental behaviour with respect to interdependence over climate change. This is because interdependence means that we have to engage with one another which inevitably leads to a dynamic notion of the concept whose meaning is constructed through such engagement (Smith et al. 2007—cited in Chap. 10). Within this construction, both common and diverse values will inevitably play a large part. The basic common value, however, is that interdependence implies respect for the knowledge of others, including a willingness to listen, share, and generate creative common solutions. Thinking about common values and how they might be promoted leads us to consider the second of the general ways in which institutionalised power and *its* underlying values may be challenged collectively through public action.

The importance of values in mediating our responses to climate change has been covered in Chap. 2. Here, we start with the prior question: where do values come from? As with everything else in our lives as social beings they are, of course, learned, reinforced, adapted or changed, drawing on many sources. For example, from:

- Others with whom we engage, especially those such as family with whom we engage a good deal.
- Written texts and audio-visual images about other lives and forms of knowledge.
- Our personal socio-economic background and circumstances.
- Our own actions and what we see going on around us which affects us either directly or indirectly.

In short, values are rooted in our lived experiences, of which they comprise a form of abstraction and generalisation. Arising from lived experience and involving a process of learning, they ultimately concern knowledge. Lived experience itself, however, is a learning process which is in turn influenced by our values. Thus, one informs the other in a circular relationship.

Sen's (1999) thesis concerning development as freedom to pursue the things we value in life (Chap. 6) tends to focus on the individual. To leverage greater power to realise what one values, however, often means collective public action with others who share those values. In this way they may act as a countervailing power to those who set climate change agendas through institutionalised authority. Thus:

- The Batwa and Ogiek communities of East Africa whose lived experiences are dominated by vulnerability—dimensions of which include structural poverty, illiteracy and displacement—value their identity as forest dwellers. It is because this identity is under threat that they ally with other minority groups internationally in an attempt to make their presence felt and overcome general feelings of powerlessness.
- The UK community campaigning against a wind farm development near their village value 'peace and quiet' in the countryside where their 'homes look out onto fields and pretty rolling hills' (Chap. 6). This common value brings them together, no matter how diverse their individual lived experiences might be in other ways.
- Charlene's (Chap. 1) 'passion for the environment and love of nature' grew from her lived experience of growing up on a farm in Zimbabwe and engaging with her father who was 'an inspiration to me'. It led her eventually to establish the NGO Environment Africa (Chaps. 1, 6, 8 and 10) which engages tirelessly on environmental matters.

- Richard (Chap. 1) similarly comes from a farming background—rain-fed agriculture in Uganda. His lived experiences of the impacts on livelihoods of increasingly erratic rainfall have influenced his values of preserving and increasing the adaptive capacity of the most vulnerable communities in his country. He also works for an environmental NGO—the Uganda Coalition for Sustainable Development.

Other people, usually from more affluent countries, have values that do not necessarily derive from the direct lived experience of climate-related and other adverse environmental impacts that have driven Charlene and Richard. Instead, a significant part of their personal lived experiences concerns their reflections on what they witness with respect to the lives of others. Here is a brief story from the 'Creative Climate' website[1] of Philippa Rowland who attended the ill-fated United Nations COP15 (Chaps. 4 and 9) that was held in Copenhagen in 2009:

> As an Australian, I went to our [Climate Action Network Australia (CANA)—a network of Australian non-governmental organisations] side event today, which had people from the South Pacific sharing stories about their islands which are going under (i.e. drowning)… They were talking about visitors coming to the island of Kiribas (officially Kiribati—comprising a group of 33 equatorial Pacific islands) and coinciding with the King Tide, so a lot of the community's actually on a house on stilts. They were on the veranda, laughing and joking, but this is serious. But the response was: 'Yes but we live with this climate change, we're adapting every moment of every day to cope with what's happening, and underneath there's a lot of grief and concern, but we're living life with joy and hoping for a better future and a safer place for our kids'. And to see these people speaking from the heart their stories and then getting up and dancing their dances was just, I was in tears, you know, I was so moved.
>
> And I think the other pivotal point for me in coming (to Copenhagen) was that Friends of the Earth brought two Nepalese people to Australia with 'The Big Melt', which is about the impact of the glaciers melting on the Himalayan region. One of them was Tenzin Dorjee Sherpa, who has climbed Mount Everest ten times, and he did it from base camp to the top in 8 h 10 min, and one reason he did that was because of the glaciers melting; there's less snow and ice around. But then he shared that his own village now has one of these glacier melt lakes above it, and old people lie in bed at night, and when it rains they wonder whether they should get up and run to the cave or whether they should stay in bed because they're waiting to see when the moraine might shift and the water might come. And I've been to Nepal twice, and I love it as a country and I love the people, and it just moves me that people that have done so little to contribute to the problem are so much copping the hard end of the stick in terms of the impact. So that fuels a commitment to come and share stories, yeah (Rowland 2009).

For Philippa and many others from affluent countries, interdependence is thus far from instrumental. It concerns fundamental human values, albeit values that are

[1]Creative Climate is a diary project of the UK Open University (OU) that records how people understand and respond to environmental change. It is available through the OU's Open Educational Resources Facility, Open Learn: http://www.open.edu/openlearn/nature-environment/the-environment/creative-climate/about-creative-climate-the-ou.

Fig. 11.1 People march during a rally against climate change. Manhattan, New York 21 September 2014 [Reuters]. http://www.euractiv.com/sections/energy/300-000-march-against-climate-change-new-york-308605#comment-1

nevertheless still rooted in lived experience. They concern justice, fairness and responsibility towards others of both future generations and the more vulnerable in the world we inhabit today. Being broad, these values transcend climate change, morphing into other global social issues of poverty and inequality, terrorism, civil conflict and human displacement.

Facilitated by social media, such values spawn social movements—to quote Wikipedia (2015), 'large, sometimes informal, groupings of individuals or organisations which focus on specific political or *social* issues. In other words, they carry out, resist or undo a *social* change' (emphases in original). Figure 11.1 illustrates a social movement group in action.

Throughout her book *This changes everything: capitalism vs the climate*, Klein (2014) espouses social movements as a countervailing power and the glimmer of light for meeting climate change and other global challenges. Thus, at one point, after reviewing positively a conference presentation by a complex systems analyst, she trumpets: 'In other words, only mass social movements can save us now. Because we know where the present (capitalist) system left unchecked is headed. We also know, I would add, how that system will deal with the reality of serial, climate-related disasters: with profiteering, and escalating barbarism to segregate the losers from the winners' (Ibid: 450).

Each of the three sentences in this quotation contains knowledge claims that are disputable. Leaving this aside, however, Klein is very clear that social movements involve countervailing values that challenge those of institutionalised power. For example:

> 'If there is a reason for social movements to exist, it is not to accept dominant values as fixed and unchangeable but to offer other ways to live – to wage, and win, a battle of world views' (Ibid: 61)… values about what we owe to one another based on our humanity (Ibid: 461)… the moral voice of the climate movement (Ibid: 463).

Social movements based on shared values that are in turn rooted in lived experiences—either directly or as witness to the generalised damage that climate change will cause, especially to the already vulnerable—undoubtedly play a significant role in public action. Playing a significant role, however, is not the same as suggesting that they represent our main, perhaps only, hope and we critique Klein's exposition of the latter on four grounds. First, she illustrates her argument with reference to the great historical social movements that achieved so much in the nineteenth and twentieth centuries, but these bear no substantive resemblance to today's climate change challenge. Second, she presents no informed vision of an overarching alternative system. Third, she does not consider the conflicting personal values that make it very difficult for many citizens to align with social movements, no matter how sympathetic they might be to their aims. Fourthly, by framing the climate challenge as a battle/war she rejects the possibility of win-win solutions. See Box 11.1 for elaboration of our critique.

Taken together, these four grounds undermine the credibility of her claims for climate change social movements. We reiterate our view that they may contribute significantly towards countervailing institutionalised power and knowledge about climate change, particularly in relation to society-wide value shifts. We argue, however, that they are not by themselves going to change the world and we should accept a more modest role which does not raise expectations too highly, where they sit among other forms of public action. Lived experiences of dashed expectations can easily turn to disillusion.

Box 11.1 The power of social movements? A critique of Naomi's Klein's *This changes everything: capitalism versus the climate*

Naomi Klein makes a strong case for social movements to bring institutionalised power to heel and force it to take strong action on climate change as a basis for tackling other enduring global challenges of poverty, inequality and social justice. A call to arms, through their actions they may trigger a complete overhaul of the capitalist economic system. In our view, however, her analysis is flawed for the following reasons.

First, she draws parallels with historical social movements which include: the abolition of slavery, the 'New Deal' that the American labour movement gained in the aftermath of the Great Depression from 1929 to the late 1930s, the national liberation movements against colonialism, the 1950s/1960s American Civil Rights Movement that achieved so much for Black equality, the 1960s second wave of feminism and the anti-apartheid movement of the 1970s and 1980s.

The parallels of these social movements with climate change activism are, however, inappropriate for deeper reasons than her noting of their mixed records in terms of tangible economic gains. Their power lay in the fact that they represented the culmination of injustices over a great many years. For example, the American Civil Rights Movement represented the culmination of a century or more of direct and crude exploitation of black people in the southern United States (even after slavery was formally abolished in

1865). Climate change on the other hand starts in the present and projects to the future which in the final analysis is still unknown despite the wealth of informed literature about what is likely to happen under global warming scenarios. And, while an argument can be made that climate change is exploitative by the world's rich of the world's poor, it is at least one step removed from the direct exploitation suffered by Black America. The latter was essentially a single issue campaign not against, but for inclusion and sharing in, the American lifestyle. This is a lifestyle that is underpinned by global capitalism which Klein explicitly bridles against as she declares that the climate change movement 'hinges on pulling off a profound and radical *economic* transformation' (Ibid: 452).

Second, while there is a strong idea in the book of what the mass movement would be *against*, as signified by the title—against capitalism—there is little to suggest what it would be *for* in terms of an overarching alternative system, although she acknowledges the need for one when she states: 'If opposition movements are to do more than burn bright and then burn out, they will need a comprehensive vision for what should emerge in the place of our failing system, as well as serious political strategies for how to achieve these goals' (Ibid: 9). Later she quotes approvingly, the Environmental Justice Movement (Chap. 2) slogan: 'system change, not climate change' (Ibid: 165).

Nor is she short of value-based aspirations for the new system as she declares the need for a powerful mass movement that weaves climate and other pressing issues 'into a coherent narrative about how to protect humanity from the ravages of both a savagely unjust economic system and a destabilised climate system' (Ibid: 7, 8), and for 'building a much more stable and equitable economic system, one that strengthens and transforms the public sphere, generates plentiful, dignified work, and radically reins in corporate greed' (Ibid: 21). She even suggests specific systems for different sectors, for example, 'combined-cycle' power plants that can switch between energy sources (Ibid: 129), transit systems accountable to their riders, water systems overseen by their users and neighbourhoods planned democratically (Ibid: 133).

Aspirations and sectoral systems, however, need an overarching institutional system or framework that sets the general rules of the game. While Klein declares war on capitalism, no alternative system is forthcoming. Nor is she enamoured by the historical alternative to capitalism, noting that 'humans have acted in this way not only under capitalist systems, but under systems that called themselves socialist as well' (Ibid: 159).

We, the authors of this book, stress here that we are not asking that she offers the detailed knowledge of a blueprint for an alternative system, but for an outline that might inform a mass social movement and help it not to burn out while it works out something better.

Our third line of critique extends the point made by Hulme (2014) in a review of Klein's book. He argues that many will not be convinced of the moral imperative to action that she calls for or, if they are, will not feel able to be part of a mass movement. He concludes: 'The world does not move together, and nor can it. Some things do not change'.

We add to this critique by asking, who is included in this mass movement? Presumably those who are seized by the moral imperative to act. Most people, however, live with a mixture of inner values, and sometimes these values will come into conflict with one another. The student discussion of personal carbon footprint that we recorded in Chap. 6 highlights such conflicts. Many people who are concerned about climate change, who understand its connections with inequity, justice and fairness, also value personal lifestyles that they have chosen for themselves and their children, and for which they work hard. They do not, therefore, keep their personal carbon footprints as low as they would like. There are limits on what they are prepared to do in the name of an interdependent humanity. As the English expression goes, they 'want their cake and to be able to eat it'. Despite this they are not without moral fibre. Nor are they a fifth column supporting corporate power, but concerned intelligent people going about their daily lives and making choices with the circumscribed freedoms that they have. What do their value dilemmas indicate for sustaining a truly mass social movement from all sides of the political spectrum? We suggest, not as much as its advocates claim.

Finally, Klein frames climate change as a 'battle of world views' (Ibid: 61) or even a war when she declares: 'Our economic system and our planetary system are now at war' (Ibid: 21). Her social movements then wage war—as in all good wars—in the name of cherished values. We critiqued this framing in Chaps. 6 and 7 on the grounds that it is deterministic, closes the door on 'win-win' responses, and demands sacrifice that many are unable to give.

11.3 Public Action, Knowledge and Engagement on Climate Change: The Broader Picture

Whether they are international alliances, social movements, NGOs, government-led focus groups, local opposition to mitigation proposals, or corporate sector influences, we see public action and engagement at every turn of climate change deliberations. Instrumental or value-led (or more usually a blend of the two), small scale or large scale, such engagements may reinforce existing knowledge through adding to it or adapting it. Conversely they may challenge existing knowledge or step aside and complement it, often through demonstration of innovative interventions.

To give some examples from previous chapters, we find under the public action umbrella:

- Engagements within the science disciplines concerning propositions and interpretations of empirical evidence, which may challenge or reinforce previous analysis (Chap. 3).
- Alliances between scientific disciplines that provide powerful story lines or axioms, such as that between physical science and classical economics: we can save the planet and create prosperity (Chap. 3).
- Engagements between the scientific disciplines which generate interdisciplinary knowledge (Chap. 4).
- Formalised engagements between local knowledge-based innovations that lead to community-based adaptation to climate change (Chap. 5).
- Formalised engagements between local knowledge, scientists and other professionals (Chap. 5).
- Oppositional engagement between institutional authority and local communities on the local impacts of mitigation schemes (Chap. 6).

Our book has done more, however, than simply record examples of the kinds of engagement that occur. It has also pointed to the boundaries that are placed around different forms of knowledge—such as the disciplinary boundaries of the sciences and socially organised denial of lived experience—making each partial and which limit engagement. Yet, beyond that, it has pointed to the potential of breaking down these boundaries through engagement, leading to enriched knowledge. Thus, we argue the potential to transcend the limitations of all knowledge boundaries through public action, via for example:

- Engagements between diverse lived experiences, essentially over constructing our interdependence and collective values across social divides, and which potentially delivers a social imagination and the power to set public agendas (Chap. 8).
- Engagements between sciences and lived experiences with respect to the meaning of impacts of climate change on everyday lives. This also generates potentially, or at least aspires to, a social imagination and a collective power to set public agendas (Chap. 8).
- Engagement of lived experiences with policy makers to establish legitimacy of interventions. This may involve becoming institutionalised and working inside the formal policy making process (Chap. 10). Alternatively, as with social movements (this chapter), it might set itself up outside, and in opposition to, what it perceives to be the inherent flaws in formal processes and the values that lie behind them. Here, without strong institutional rules, the engagement process is inherently more unruly. Both have their strengths and weaknesses. Both have more or less validity. Neither provides a complete approach.

If this feels like a theory of muddling through the climate challenge, so be it. We do not, however, make a fetish of it. We have no underlying assumption of deliberately seeking consensus or disagreement, but note that both will occur, probably simultaneously with consensus on some matters and discord on others, and both may be open to critical analysis. We do argue, however, that if this is the way it is, let's make the most of it and not abrogate our responsibilities as citizens to engage.

In this, we should never forget the potential role of *informed* citizens operating individually or collectively in defining public policies and actions. Citizen public action is more than 'green' behaviour in terms of, for example, doing our bit to reduce carbon footprints. It is also more than our right in many countries to elect and de-elect our governments, important as that is. (In any case, at an international scale, a world government that is democratically accountable is not even on the radar.) Nor does it necessarily concern our ability to mount 10, or even 100, demonstrations relating to a United Nations climate change COP. It does concern, however, our abilities to apply individual and collective pressure through a combination of working with, and where necessary confronting, governments and their international manifestations and demonstrating alternatives.

In this regard we should not demand too much of the annual United Nations COPs as something external to us. They are part of the process but a form of responsibility abrogation occurs when we raise our expectations on what they might deliver, thus setting them up as straw people who fall down when they fall short.

We stress here the qualifying adjective 'informed' which we don't restrict to citizen understanding of the science of climate change and its likely impacts, nor of the social science of understanding socio-economic impacts. Such understandings are undoubtedly necessary to be 'informed' but they are not sufficient. Knowing the 'facts' is neither enough to change personal lifestyles nor to change policy. This is because the 'facts' might contribute to our knowledge, but they do not represent that knowledge. We reflect on 'facts' and assimilate them through our personal and collective experiences, experiences which are influenced by our histories, our material and social circumstances, our ways of life and mindsets, and so on.

Thus, by 'informed', we do not mean an end state of knowing the 'facts' but a process of 'informing and being informed' through engagement with others. That for us is the true measure of citizenship, and one which can help meet the climate change challenge.

It seems that at the heart of policy disputes about climate change, which take place at all scales, from micro- to macro-, are disputes about knowledge which then becomes a political tool and focal point for public action. Thus, we can conclude that climate change knowledge involves multiple agents with public and private interests. As a result it is forever emergent and potentially conflictual, but where, at certain points, accommodations can be found. From these points a policy that at least most agents are prepared to live with is provisionally accepted, if never completely agreed.

Also, we reiterate, the public action approach reveals that knowledge construction for policy and intervention is never easy politically. More than likely, it is a somewhat unruly, political process of disputed scientific evidence and oppositional lived experiences, although this need not always be the case Unruly processes (even where the degree of unruliness is quite minor) tend to lead to polarisation of conflicting interests, where one party 'wins' and another 'loses'. Ultimately it is knowledge itself that becomes the loser because the opportunity to put diversity to constructive use has been lost. Can the process be made less unruly, therefore, by incorporating and institutionalising lived experiences at an early stage and consciously avoiding polarisation before it is too late? Possibly, as Chap. 10 demonstrates, but two fundamental principles need to be institutionalised.

The first involves a shift from viewing knowledge, whether scientific or lived experiential, as something to extract and represent in the policy process, to viewing it as a process of generation through engagement between diverse scientific and lay actors. In other words, such a process would involve accepting that different lived experiences need to engage actively with the sciences as well as one another, and providing the space for this to happen.

Second, it is important that the experiences of a wide variety of citizens and social groups are brought into the debate alongside current scientific knowledge and its conflicts. Yes the science has to be robust, but so too has political acceptability and legitimacy of policy and intervention proposals.

The prize of a transboundary social imagination awaits, that enables citizens—and note that scientists and other climate change professionals are also citizens—to establish the public agenda surrounding climate change.

References

Abbott, F. (2015). *Personal communication.*

Chivers, D. (2015). Whose renewable energy? *New Internationalist.* March 2015, 12–17.

Hulme, M. (2014). *Does this climate narrative really change anything?* Review of Naomi Klein's 'This changes everything: Capitalism vs. the climate'. London: New Scientist Magazine (23 September 2014).

Klein, N. (2014). *This changes everything: Capitalism vs the climate.* London: Penguin.

Mills, C. W. (1959/2000). *The sociological imagination.* Oxford: Oxford University Press.

Norgaard, K. M. (2011). *Living in denial: Climate change, emotions and everyday life.* Cambridge, Massachusetts: The MIT Press.

Rowland, P. (2009). *Philippa's Story.* From 'Creative Climate', Milton Keynes: The Open University: http://www.open.ac.uk/openlearn/nature-environment/the-environment/creative-climate/explore-the-stories/community/under-dripping-glacier. Accessed February 26, 2015.

Sen, A. K. (1999). *Development as freedom.* Oxford: Oxford University Press.

Smith, J., Clark, N., & Yusoff, K. (2007). Interdependence. *Geography Compass, 1*(3), 340–359. doi:10.1111/j.1749-8198.2007.00015.x.

Stern, N. (2010). *Blueprint for a safer planet: How we can save the world and create prosperity.* London: Vintage Books.

Wikipedia. (2015). *Social Movements definition.* http://en.wikipedia.org/wiki/Social_movement. Accessed February 26, 2015.

MIX
Papier aus verantwortungsvollen Quellen
Paper from responsible sources
FSC® C105338

If you have any concerns about our products,
you can contact us on
ProductSafety@springernature.com

In case Publisher is established outside the EU,
the EU authorized representative is:
Springer Nature Customer Service Center GmbH
Europaplatz 3, 69115 Heidelberg, Germany

Printed by Libri Plureos GmbH
in Hamburg, Germany